Safety Management in Construction and Industry

David Goldsmith
President, Goldsmith Engineering Services

McGraw-Hill Book Company

New York   St. Louis   San Francisco   Auckland   Bogotá
Hamburg   London   Madrid   Mexico   Milan
Montreal   New Delhi   Panama   Paris   São Paulo
Singapore   Sydney   Tokyo   Toronto

Library of Congress Cataloging-in-Publication Data

Goldsmith, David, date
    Safety and management in construction and industry.

    Includes index.
    1. Industrial safety. 2. Construction industry—
Safety measures. I. Title.
HD7262.G58   1987   363.1′19624   86-27733
ISBN 0-07-023677-1

1234567890   DOC/DOC   89210987

ISBN 0-07-023677-1

*The editors for this book were Nadine M. Post and Marci Nugent,
the designer was Naomi Auerbach, and the production supervisor was
Dianne L. Walber. It was set in Century Schoolbook by The William Byrd Press, Inc.*

Printed and bound by R. R. Donnelley & Sons Company.

Contents

Preface   v
Introduction   vii

Chapter 1. Safety Control: The System Basics   1

Delegation of a Safety Supervisor   2
Job-Site Personnel Orientation   4
Delegation of Craft Spokespersons   8
Toolbox Meetings   9
Safety Committee Organization   10
Safety Committee Duties   12
Organization of the Safety Program   13
Reporting System   16
Sample Forms   18

Chapter 2. Elimination of Potential Safety Hazards   33

Daily Job-Site Surveys   33
Weekly Job-Site Inspections   35
Monthly Job-Site Inspections   35
Historical Data Hazard Analysis   36
Planning and Site Hazard Analysis   37
Job Hazard Analysis   41

Chapter 3. Planning and Scheduling for a Safety Program   45

Planning a Safety Program   45
Planning for Mobilization and for the Subcontractors' Arriving on Site   47
Project Construction   49
Commissioning or Plant Start-up   51
Demobilization   55

Chapter 4. The Safety Program "Up and Running"   57

Chapter 5. A Few Interesting Facts   73

Common Causes of Accidents   73
Sources of Accidents   76
Safety Aids   77
Safety Hazards   79

Carelessness   83
Drugs and Alcohol Abuse   85
The Federal Hazard Communication and "Right-to-Know" Laws   88

Chapter 6. Self-Evaluation of the Safety Program   91

Review of Site Inspections   91
Review of Committee Meeting Minutes   92
Review of Orientation Sheets   92
Review of All Accidents   93
Miscellaneous Reviews   93
Conclusions   95

Chapter 7. The Safety Program Audit   97

Procedure Philosophy   98
Inspection and Audit Checklist Forms   98
Arrival-on-Site and Entrance Interviews   98
On-Site, In-the-Field Procedures   99
Site Inspection Evaluation and Exit Interviews   104
Site Evaluation Report   106

Appendix   A Typical Audit Inspection Checklist   107

References   115

Index   116

Preface

Heavy-manufacturing industries and industrial construction have long been plagued by high accident rates, and after several years of holding the rate in check and in some cases reducing it, the rates are now increasing again. Federal and state governments, trade unions, and insurance companies have been extremely concerned about this state of affairs for a long time. But because of the nature of these industries and the diversification of each area of work and each task performed, it has been extremely difficult to regulate and standardize methods of control for monitoring safety practices and procedures.

Government, unions, and insurers have spent a great deal of time and effort attempting to provide legislation and rules and regulations to help reduce the large losses of life and limbs and the high number of "lost-work days." But all the legislation and regulation in the world will not reduce accident rates unless craftspeople and management take positive action to put these rules into practice in their everyday activities and really start to care about safety in their work environment. Management and labor must step forward together with a solid plan to *help themselves* reduce the appalling number of work-related injuries. It has been proven that when both groups work together, safe work conditions can be achieved.

The purpose of this book is to set forth an effective, economical system of safety management, safety control, and hazard analysis suitable for use in any construction project or heavy industrial plant. Further, it can easily be supplemented with other records and systems where a particular type of plant has unique problems or unique activities which are not included in this illustrated system.

The system outlined in the following chapters is simple, basic, effective, proven, and, most important, economical so that any project, no matter how tight the financial situation, can use it without expending a great deal of extra cost and effort. Staff additions are not necessary, and, if it is followed diligently, the system offers the prospect of saving money, increasing production, and, most important, reducing expensive lost-time accidents.

These same heavy-manufacturing (e.g., iron, steel, petrochemical, and power) and construction industries encompass a highly talented nucleus of companies and professionals, scholars, engineers, and craftspeople. With this simple program and the expertise and efforts of the professionals on site, it is surely possible to significantly reduce accidents and to rejuvenate morale to the point that these industries can return to being the pride of the nation and that going to work for all industrial personnel can be a pleasure instead of a daily lottery.

The cartoon characters and the skits that appear in Chapter 5 are reproduced with the kind permission of Bethlehem Steel Corporation. Bethlehem Steel republished their booklet in which the cartoons originally appeared, *SAFETY CODE*, for the Steel Construction Division in 1962. The booklet is now regarded as a collector's piece. It was the only one of its kind and was far ahead of its time, showing that some employers in those days really cared about their employees. It had a significant impact on their erection department and helped to stem the tide in the high accident rate among steel erectors. It is interesting to note that Bethlehem Steel was then, and is now, at the forefront of safety in the industrial workplace. They are strongly committed to a safe work environment.

The writing of this book would not have been possible without the help and encouragement of many people. These people are very well versed in industrial safety with both heavy theoretical and practical experience. I owe them all a debt of gratitude for their unfailing advice and support.

My wife, Dawn, and my father, Eric, both "proofreaders extraordinaire," to name but one of their collective talents
Bill Burdick, safety consultant with Indiana University, Pennsylvania
Jim Lapping, safety director for the AFL-CIO, Washington, D.C.
Jim O'Brien, manager of safety, Dravo Corporation
Lew Weidensaul, technical director, Liberty Mutual Insurance Company
Tom Doherty, OSHA Regional Office, Harrisburg, Pennsylvania
David Bell, OSHA construction advisor, Washington, D.C.
Peggy Richardson, OSHA, Manager of VIP Programs, Washington, D.C.

And finally a big thank you to Lisa Kyper for her skill at translating my scrawl into a relatively intelligible manuscript.

David Goldsmith

Introduction

Everybody on the site, from manager to coffee vendor, should participate in job-site safety activities. And equally important, safety activities must have committed corporate support.

Many companies who have not already established programs want to start safety programs, but they feel that the cost would be prohibitive or that the time or the wherewithall to develop a program does not exist.

This book provides a simple system that can be implemented without an inordinate outlay of money or time. This system satisfies all the legislative demands for even the smallest of operations, and, if necessary, it can be expanded as required. The book has been written so that a management group or a union group or both can, with a minimum of change, apply the procedures to their own needs without copious paperwork or wastage of time. Incorporating all the frills, only a few hours per week are necessary to keep the site safe by systems control rather than haphazard crisis control or crisis management, as is becoming all too prevalent.

Any safety program should be developed at the bid stage of a construction project, but obviously such a program can be picked up and used at any point along the way. The safety program described in this book can be put into place at any time during the life of a project.

The system delineates responsibilities and accountabilities. It outlines procedures for eliminating hazards and identifying potential hazards before they become contributing factors in unfortunate accidents. In addition, it illustrates how normal, common management tools, particularly forward planning and scheduling, can be used to produce safety schedules from the initial bid stage of a prospective project through contract completion and plant operation. It also provides a simple procedure for "activity hazard analysis" similar to that being used in conventional predictable industrial work (job hazard analysis, or JHA). Finally, it provides a procedure for self-evaluation by the people who use the system to evaluate how well the program is working. The self-evaluation reveals weaknesses—that is, particular functions within the program that are not up to the standards expected or anticipated—that can then be improved upon or eliminated.

It is not suggested for one minute that this program is the only answer to the subject of safety on the job site, but this program can be used in part or in total to help individuals and organizations establish safety standards on any project on which they wish to use it.

For little additional outlay in time and money, there is an opportunity to show magnificent returns—a lower accident rate and increased profit. So, for whatever reason an organization decides to use a safety program, it should be remembered that the program in this book, together with dedication, understanding, tenacity, and conviction, will, in large measure, remove most of the potential hazards on site.

And if this program is implemented on only one site and prevents one accident, then the program will certainly have been worthwhile.

Safety Control:
The System Basics

For any industrial and construction project in the United States, certain federal laws exist that establish the necessary requirements for a minimum standard of safety and health to be achieved. These standards are listed in the *OSHA Safety and Health Industry Standards for General Industry and the Construction Industry.*

The following is an abbreviated* list of the OSHA minimum requirements of a safety and health program for contractors and subcontractors on every job site:

1. The employer shall initiate and maintain a job-site safety and health program.

2. No construction or industrial worker will be required to perform work under conditions that are either unsanitary, hazardous, or dangerous to her or his safety or health.

3. Frequent and regular inspections of job-site materials and equipment shall be conducted by competent persons designated by the employer.

4. The use of any machines, tools, or equipment which are not in compliance with the safety standards is prohibited.

5. All equipment that does not comply with the safety standards must be identified by tagging and the controls locked to prevent use, or they must be removed from the job site.

---

* The word "abbreviated" does not mean there are more requirements, simply that all the legal jargon and auxiliary whys and wherefores have been omitted to just give the basic facts.

6. Only employees qualified by training or experience may operate equipment.

7. Every employee shall be instructed in the safety and health regulations applicable to his or her work.

8. The OSHA poster must be displayed, and copies of the act and OSHA rules must be available to employees upon request.

9. A record of all lost-time injuries must be kept and posted at the job site.

10. All fatalities and accidents which result in the hospitalization of five or more workers must be reported to the OSHA area director.

11. The name and location of the nearest medical facilities must be prominently posted at the job site.

12. Every employee must be instructed in the recognition and avoidance of unsafe conditions.

13. An employee representative must be given the opportunity to participate in the OSHA inspector's opening and closing conferences and all informal conferences.

14. All workers must have access to exposure and medical records.

The standards in the preceding list were all extracted from 29 CFR 1903, 1904, and 1926, and many more standards will be added in the future until we in construction and industry help ourselves by reducing existing accident rates on a national scale.

Our obligations in establishing safe industrial and construction job sites have been clearly spelled out. Now we must start implementing a comprehensive system to incorporate all this legislation. Remember, it is very easy to do the bare minimum in anything. But eventually somebody gets careless, and the bare minimum is not good enough anymore. This carelessness could cost somebody a great deal of money and, even more disturbing, somebody somewhere some serious physical damage.

Therefore, why not start with a complete program which covers all the requirements very satisfactorily, which does not take any significant increase in time, money, or effort, but which ensures that everybody becomes involved to some extent in their own well-being on the job site? So let's begin the program by realizing that the basic thread is communication, and without it there will not be a good safety program.

Delegation of a Safety Supervisor

Many titles exist for management personnel who are responsible for job-site safety. The following is a list of some of these titles:

1. Safety engineer
2. Safety officer
3. Loss-prevention coordinator
4. Hazard control manager
5. Job-site safety analyst
6. Work protection supervisor
7. Accident prevention superintendent

The responsibilities for job-site safety are usually delegated to a manager in addition to his or her primary job function. However, because of conflict of interest, neither the site manager nor the project manager should be the safety supervisor unless she or he is the only management person on site. It has recently become common practice that the site engineering department appoint the safety supervisor since the site engineers are constantly working with safety considerations in the course of their normal operations. In addition, the progression to safety supervision is natural since the site engineering department is usually represented on site from project start to finish.

Responsibilities of the safety supervisor

1. He or she is heavily involved in site planning and scheduling.
2. He or she is heavily involved in site construction and its problems (which, of course, usually are the "mother" of safety hazards).
3. He or she should be familiar with the bid documents, the budget, and the project installation.
4. He or she must objectively and continuously evaluate work areas and procedures.
5. He or she must be on site intermittently and as often as necessary to make realistic assessments of changing conditions and new situations which breed the opportunities for hazards to exist.

It is clear from the preceding list of responsibilities that the person who is delegated as safety supervisor be first and foremost *competent and capable of dealing with identification and correction of existing or predictable hazards*. Further, that person will have more success if he or she is an able administrator. These criteria, of course, limit the field of choice, but it is very important that the person with the best chance of success be chosen. Such qualifications can best be assessed with established personnel selection procedures.

Having been delegated this responsibility, this person becomes totally responsible for site safety activities. However, these activities needn't take up all his or her time. For example, on a 350,000-manhour project over a 2-year period, it will take only this one competent person approximately 3 to 8 hours per week to administer a successful safety program. Therefore, the duties this person assumes as safety supervisor can be very easily fitted in with his or her primary duties as, e.g., field engineer or superintendent.

Duties of the safety supervisor

(Note that the following duties are not listed in any order of priority.)

1. Develop and distribute information for toolbox safety topics.
2. Organize, establish, and maintain complete files for the safety department.
3. Ensure that all the reporting paperwork (forms and reports) is completed, processed, and distributed.
4. Coordinate and chair the weekly safety committee meetings.
5. Organize and participate in monthly site inspections and develop daily inspection criteria.
6. Implement and distribute all company safety procedures and information.
7. Liaise between site and company corporate safety department, subcontractors, and other organizations with reference to safety topics.
8. Coordinate the safety program with the operations and construction activities.
9. Arrange site safety and hazard-analysis training.
10. Evaluate all accidents and near accidents on site.
11. Continually monitor the safety program for functionability and improvement.
12. Perform or supervise all personnel job-site orientations.
13. Prepare and put into place an emergency evacuation plan and an emergency injury plan.
14. Institute a suggestion box for suggestions and complaints, and review them on a regular basis.

Job-Site Personnel Orientation

This duty consists of introducing *all* employees (salaried and hourly people) to the company and the job site and the safety policies, rules, and regulations of both. Carefully and concisely, the safety supervisor should explain all standard company safety rules, the safety program, and any special regulations peculiar to that particular project (for example, one job may be surrounded by railway tracks, in which case special signals must be heard, understood, and obeyed.)

This orientation can be done in a speech to a few employees or with a short video, if available, for a large group of employees. Both methods are equally positive, and choosing one or the other is purely a matter of company management preference. This initiation should be recorded and detailed on a form similar to Form 1.1, which is presented at the end of this chapter.

It is imperative that principal site personnel of any subcontractor on site also be indoctrinated, so that they in turn may incorporate all the rules and regulations of the general contractor or supervising contract group when giving their own orientation presentation to their own craftspersons. Each of these orientation sessions should also be documented with copies forwarded to the general contractor. The ultimate purpose of recording the orientation information and seminar is that such documentation becomes the equivalent of a contract of employment for the protection of both parties—management and labor. It is a record of the agreement between both parties which can be referred to should either side renege on their obligations.

The following checklist is useful for preparing a safety program orientation:

1. Briefly describe the company and its history.

2. Incorporate a corporate-level statement from the company in which the company's commitment to maintaining a safe job site is put forth (see Figure 1.1).

3. Provide a list of the principles of safety as related to the particular industry (Figure 1.2 is an example of such a handout.)

4. Explain the safety requirements of each individual trade activity, relative to the assembled groups (Figures 1.3 through 1.6).

The principles of safety and health management are based on a strong commitment by the company management to promote protection of personnel and property.

The company considers no discipline of the company operations more important than accident prevention.

Planning for safety will start during the project bid cycle and continue through design, purchasing, and all other functions of the services provided by the company for its clients.

Planning to minimize potential hazardous conditions and incorporate OSHA requirements is the direct responsibility of the operating management. All supervisors will accept responsibility for the prevention of accidents in the areas of his or her responsibilities.

Figure 1.1 Corporate policy statement on safety and health.

Floor Openings and Floor Holes

Floor openings and floor holes must be covered or protected by a guardrail.

Covers must be cleated and constructed of minimum ¾-inch plywood or equivalent.

Guardrails must have a top rail 42 inches high, a midrail 21 inches high, and a toeboard. Top rails must be capable of withstanding 200 pounds with minimum deflection.

Wall Openings

Whenever an opening is at least 30 inches high and 18 inches wide, and there is a drop of more than 4 feet, and the bottom of the opening is less than 3 feet above the working surface, the wall opening in question must be protected by a standard guardrail as described above.

Scaffolds

1. Work Platforms Which Are More than 10 Feet High

   Mandatory requirements:
   Guardrails (consisting of top rail and midrail): The top rail must be 42 inches above work platform constructed of 2- by 4-inch timber or equivalent. The midrail must be approximately 21 inches above the work platform. The top rail must be capable of withstanding a thrust of 200 pounds with minimum deflection.
   Supports: Spaces between the guardrail and the supports are not to exceed 8-foot centers.
   Toeboards must be a minimum of 4 inches in height continuously around the platform.

2. Work Platforms between 4 and 10 Feet in Height

   Mandatory requirements for work platforms which have a minimum width in either direction of less than 45 inches:
   Guardrail or top rail must be 42 inches high above the work platform and constructed of 2- by 4-inch timber or equivalent. (In this case, only a top rail is necessary.)
   Supports: Spaces between the guardrail supports are not to exceed 8-inch centers.

Figure 1.2 Principles of safety: Examples of special job-site rules.

5. Explain the particulars of the specific project, and also explain the rules and regulations unique to that particular project. Every industry and construction project has many peculiarities; thus, while the principles may be the same, the priorities may be very different.

6. Explain the emergency evacuation plan procedures.

7. Explain the emergency treatment facility procedures.

8. Explain the fire plan procedures.

9. Identify employees with previous safety administration experience, including first aid, cardiopulmonary resuscitation (CPR), and toolbox training.

10. Thoroughly explain the job-site safety program, its advantages, and its requirements for each worker. Demand participation.

11. Refer to the ratified agreements between each respective union and management which were delineated at the prejob meeting,

---

1. At the start of each shift, check brakes, steering, indicators, warning lights and devices, and all other equipment necessary for the safe functioning of the machines you are driving.

2. When driving, you must always face the direction of travel.

3. Always make complete stops when approaching doors, corners, exits, stop signs, etc.

4. Sound a warning with the horn at all blind spots to other traffic or personnel.

5. Site speed limit is 15 miles per hour maximum.

6. Do not drive over objects on the ground.

7. Keep an adequate distance but a minimum of 20-feet distance between vehicles when on site.

8. A reversing siren is mandatory on every commercial vehicle. It is your responsibility to ensure that it is working or to report that it is not working in writing to your supervisor. The vehicle will not be used until its siren is returned to working order.

9. Ensure that a working fire extinguisher is installed in your vehicle, or report it to your supervisor if there is not one in your vehicle.

10. Make sure all loads are tied down to the bed regardless of how far the load is going.

11. Don't run your engine in a confined space.

12. Don't remove the radiator cap until (a) the cap is first relieved by "cracking" and (b) until the radiator is completely cool.

13. If the vehicle is equipped with seat belts, use them.

14. Never leave your truck running when you are not in the cab.

15. Never allow or participate in horseplay near your vehicle.

Figure 1.3 Craft training card: Truck drivers.

including, it is hoped, joint labor-management participation in the safety program.

It is quite amazing how an employee's first impression of the job site and the company are guided by this job-site orientation. If the orientation is done properly and with enthusiasm, the employee will leave with a good positive feeling, which, providing she or he follows the program guidelines, will stay with her or him throughout the whole job. Quite unconsciously, the employee will give more of herself or himself to the job because she or he is impressed and happy.

If, on the other hand, the orientation is not positive but is instead blunt or objectionable in the manner in which the information is presented, then the employees will react negatively, and they may not care about anything—least of all the safety program—except earning a dollar.

This video or verbal orientation is critical to project success and to safety program adherence. Do it well—it will pay big dividends.

Delegation of Craft Spokespersons

Each craft should elect a spokesperson as that specific trade arrives on site. This person may be the union representative, the general fore-

1. Whenever possible, avoid standing on machinery.
2. Loose-fitting clothing should not be worn when working close to moving machinery parts.
3. When refilling a fuel tank, use safety canisters with proper spouts or filling funnels to prevent fuel spills. Avoid spilling fuel on the hot exhaust pipe. However, if it is unavoidable that some fuel will spill on the exhaust pipe, wait until the engine cools down.
4. When a crane boom needs to be greased or oiled, the boom should be lowered to the ground before performing these tasks.
5. When it is necessary to remove guards to grease or oil a piece of machinery, make sure the machine is stopped and isolated before removing the guard. Make sure the guard is replaced before the machine is restarted.
6. Prevent machinery from dripping oil or grease since leakage not only provides a hazard capable of causing an accident but reduces the efficiency of the machine.
7. Report all unsafe mechanical conditions to the operator of the machine and to a responsible supervisor.

Figure 1.4 Craft training card: Operating engineers.

1. Store materials in an orderly manner for your own safety as well as your workmate's.
2. Use heavy-duty gloves when handling coarse or rough materials.
3. Make sure your body is always well-covered.
4. Since concrete products and additives can cause painful burns, use protection, including barrier creams, whenever possible. Always wash frequently.
5. Before handling discarded lumber, inspect it for nails. To avoid dangerous skin punctures, those nails that can't be pulled out should be bent over.
6. When making a lift, make sure your legs are bent and your back is straight; then straighten your legs. This will prevent back injuries.
7. Wear goggles when using hand tools.
8. Be alert when using compressed air to avoid its contact with the rest of the work force.
9. If in doubt, ask!

Figure 1.5 Craft training card: Laborers.

1. Do not perform electrical work while power is on to a circuit that could be disconnected.
2. Use temporary lighting wherever practical to assure that there is adequate light in working areas.
3. When pulling cable through conduit, it is very important that you make sure all adjacent power is completely isolated from the pulling team at either end.
4. Make sure a job is safe before walking away.
5. You should construct all temporary installations with the same degree of safety and quality of work as you would a permanent system.
6. Verify that all powered equipment is properly grounded before use.
7. Do not perform outside electrical work in the rain.

Figure 1.6 Craft training card: Electricians.

man, or a person particularly able, interested and/or experienced in safety programs. When appointing the spokesperson, the craft and site management should collaborate with the business agent to achieve site harmony. As a footnote and purely for the sake of passing on proven ideas for appraisal or cogitation: It has been found effective in this type of program to use the general foreman of the particular trade initially as the spokesperson since doing so gives the rest of the group a chance to organize, orient, and get to know their coworkers. Then, after an appropriate amount of time, the foreman and craft personnel may wish to delegate the committee representatives' duties to another person within that craft. However, it is also absolutely necessary to ensure that the spokesperson be competent in safety and health awareness. He or she must (1) be able to identify existing and predictable hazards and (2) be authorized to ensure that prompt measures to correct those hazards are taken by the responsible supervisor.

Therefore, of course, it is very important that while management is totally responsible for site safety and the health program and its management, the craftperson delegated to serve on that committee must be authorized on behalf of the committee to ensure that prompt corrective action is taken in his or her own area of responsibility, wherever and whenever necessary. The committee member must be totally supported by the safety committee at all times. In that way, the committee will be seen to act as a joint labor-management body in unison regardless of any circumstances which may arise, and confidence from all parties will increase in the program and its implementation.

There are going to be times when there are pressures of schedule and financial restraints which will tend to conflict with the safety program, and these are the times when safety concerns and the program should be most unswervingly enforced. This perseverance takes courage from everybody, and if the committee members are in unison and have the support of corporate management, it becomes a much easier task. Site management then will have to support a decision on that issue, and theoretically the unsafe act will be curtailed *before* it happens.

Toolbox Meetings

Toolbox meetings should be held weekly at a convenient time, that is, at a time that is least disruptive to the job site. The safety spokesperson for each craft should address a gathering of all the members of that particular craft; however, if the size of the crew is prohibitively large, smaller groups should be assembled.

The craft safety spokesperson also represents his or her craft on the joint managment-labor safety committee, which is elaborated upon later in this book. The most important discussion topics for these weekly 10- to 15-minute sessions are listed below. Other topics, of course, may be added as required or necessary.

1. Any outstanding business from the previous meeting should be brought out and concluded if possible.

2. Those safety topics relevant to the continued upgrading of job-site safety training and hazard analysis should be discussed. Written materials for this discussion can be obtained from the company, from the company's insurance carrier, or from OSHA upon request.

3. Projected activities for the week should be reviewed. Hazard recognition awareness and control should be emphasized. This discussion should focus on learning how to recognize hazards, how to be aware of them when working, and how to control potentially dangerous events going on around the work area.

4. Accidents or near accidents should be reviewed.

5. A report should be given on the previous safety committee meeting.

6. Any relevant job-site information should be passed on.

7. Any new safety considerations for the job site should be presented.

The gist of all the basic points discussed at the meeting, together with a list of attendees, should be recorded, dated, and filed. A typical example of a form that can be used for this purpose is presented at the end of this chapter (Form 1.2).

Safety Committee Organization

The safety committee is made up of members of the site company management team, the craft representatives already selected by their work forces, and representatives of any subcontractors on site. Initially, this committee was envisioned to be 50:50 percent labor-management participation. However, this makeup is purely a matter of preference of the people involved.

It is not necessary to have equal proportions of labor and management participating on a safety committee; in fact, larger proportions of labor usually encourage greater participation. Experience with different proportions of labor participation has demonstrated that more open and frank discussions and more positive meetings and actions result when the larger proportion comes from the labor forces. Note that it is, however, necessary to include a minimum of 50 percent craft employees as part of any safety committee.

The most effective, competent personnel to be involved from management are the general superintendent and field engineers or craft superintendents for each discipline current with the project status (critical path), and from labor each current craft representative and subcontractor representative. Generally, this selection works very well with each discipline rotating as the project status changes.

The workings of the committee will be described a little later in this book; here it is important to discuss the makeup and flexibility of the committee. The reader should be aware that the makeup of the committee is in keeping with construction procedures and is very much a part of the whole philosophy of modern-day construction hazard analysis.

For the purposes of this book, construction can be divided into four major phases of work. These encompass many different crafts and materials but nevertheless also clearly categorize the types of hazard exposures which face each discipline. These groups also offer a definitive path to rotate and regulate to a manageable size the members of the safety committee.

1. *Civil.* Site preparation, mobilization, excavation, pile driving, reinforced concrete, concrete foundations, and underground conduit all fall within this group. Therefore, carpenters, laborers, ironworkers, cement finishers, operating engineers and electricians all should be involved in the safety committee during this phase, together with any subcontractor representatives.

2. *Structural.* Structural steel, sheeting, roofing, and structural concrete all fall within this group. Therefore, carpenters, laborers, ironworkers, and operating engineers are the major participants, and, they, together with any subcontractor representatives, should be involved in the safety committee during this phase.

3. *Mechanical.* Any major machinery and vessel setting, auxiliary equipment, conduit and cable installation, fittings, services, and piping installation are included in this group. Some of the trades involved are millwrights, boilermakers, pipe fitters, plumbers, and electricians, and they, together with any specialty subcontractors, should be represented on the safety committee during this phase of construction.

4. *Finishing.* This group includes architectural work, outhouses, blockwork, coupling-up equipment, pulling and terminating electrical and instrumentation cable, machinery installation, piping, and commissioning. Therefore, the major trades involved are electricians, millwrights, pipe fitters, masons, and carpenters together with all the various commissioning technicians and experts and any remaining on-site subcontractors. Representation on the safety committee should again be broad and should always include the

testing and commissioning coordinator. *This period is just as critical as the others, and in some cases more so!*

As you will see from these groups, it is possible to have a good mixture of members for the safety committee, providing tradespeople who work all through the project and some who work during parts of it, who can replace each other as the project emphasis changes through each phase of construction.

Safety Committee Duties

1. Delegate daily, weekly, and monthly site safety inspections to specific committee members who should be accompanied by craft personnel. Ensure that all these daily, weekly, and monthly site inspections and the required reporting actually take place and review the reports.

2. Ensure that each party responsible for safety on the job site performs his or her duties.

3. Meet weekly.

4. Discuss generally and specifically the following week's construction activity, and organize any special safety actions or requirements.

5. Arrange for ongoing personnel training on site in such topics as cardiopulmonary resuscitation and first aid. Upgrade hazard recognition courses, and make video training programs available to personnel.

6. Review and comment on accidents and near accidents.

7. Review and fine-tune all contingency plans such as the emergency evacuation plan.

8. Distribute accumulated safety information at toolbox meetings.

9. Try continuously to evaluate, upgrade, and improve the safety program on the site.

10. Review work force complaints and suggestion box information.

It can be seen from the preceding list of duties that the safety committee members are the nucleus of the whole safety effort. Thus by delegating responsibility of all safety problems and actions to this group, the whole area of hazard control and analysis can be encompassed and successfully managed. When things do not go as they should, then the safety committee is able to convene, appraise the situation, regroup, and hopefully eradicate the problem, or ammend the system, *before* the oversight causes something more serious.

However, if the committee (or the safety coordinator) needs help or guidance, then the site manager should become involved since he or she is, of course, ultimately responsible for the site and its safety success or failure.

Organization of the Safety Program

At this point the committee members are fully aware of their responsibilities, and they are united as a team.

The committee has a purpose—to plan a safe work environment and reduce the opportunities for accidents to happen.

The site manager knows who does what and who is responsible for each unit, which means there are supposedly no gaps in the administration or execution of the safety program. In that way, nothing should get forgotten or missed.

The organization charts in Figures 1.7 and 1.8 show how the safety department fits into any site organization and whom it reports to.

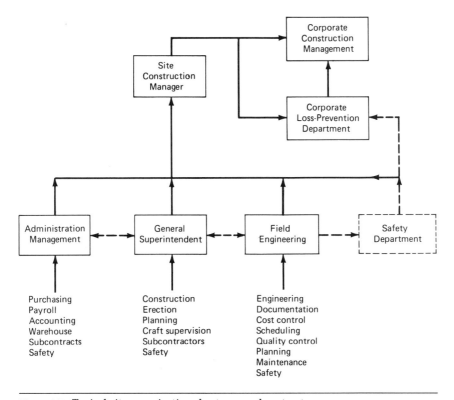

Figure 1.7 Typical site organization chart, general contractor.

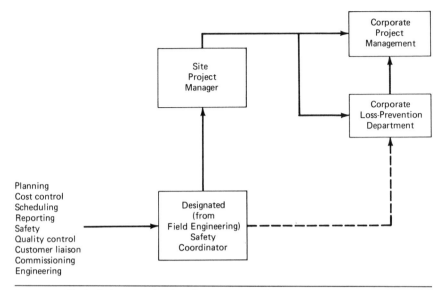

Figure 1.8 Typical site organization chart, construction-project manager.

Further, Figure 1.9 shows the makeup of the safety committee and the way the members can rotate during the various phases of construction. This model is intended to be basic and can be expanded or reorganized ad infinitum. For example, some projects require a full-time safety coordinator, and some of the megabuck projects require one safety coordinator per area. However, this model is really trying to set a minimum standard to help the small general contractor improve his or her safety program as well as serving as an example for the well-organized general contractor or construction manager, who may pick up some useful ideas.

The basic philosophy, however, is that it does not matter how many people are involved in the safety program so long as all the facets of that program are delegated to responsible people.

If the project is small, it may be that only two, three, or four persons need to be heavily involved. Of course, if the project is larger, then obviously more people will have to be involved.

The organization of duties and personnel is entirely up to the safety coordinator, the site manager, and corporate management and its policies. As long as it is realized that the most critical hands-on functions of the safety program are the site inspections and hazard analysis, the remainder of the program is easy to accomplish at odd times during the working week by one reasonably efficient administrator. Much of the paperwork is so repetitive that a capable engineering or site clerk can do some portion of it.

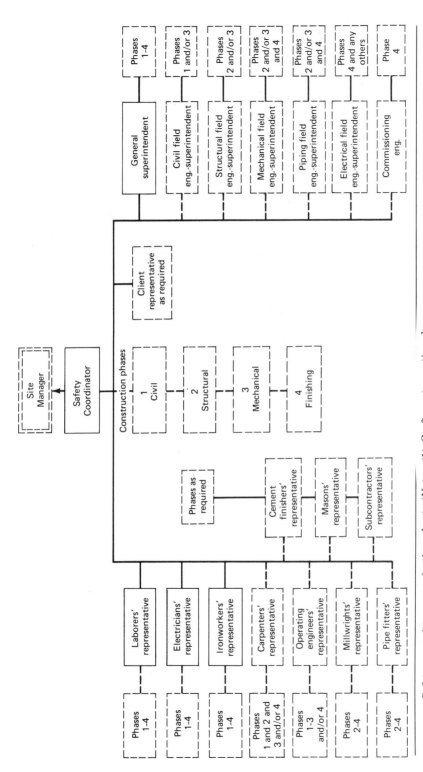

Figure 1.9 Safety committee organization chart. (*Notes*: (1) Crafts not mentioned are represented by subcontractors. (2) Broken-line boxes indicate rotating members and phases of construction. (3) The site manager does not normally participate.)

With these activities being administered on an independent basis, the safety committee can cover its agenda in the meetings swiftly without laboring on any detail for too long. Meeting documentation also ensures that even when a particular member misses the meeting occasionally because of more pressing needs, a concrete pour that cannot be stopped, for example, the meeting does not suffer through lack of continuity. Additionally, the smaller the group of members within reason, the less interruptive their meetings will be to the project production schedule, which must also be considered in establishing the safety program. There must be give and take by both groups for the program to work effectively and in harmony.

Finally, an additional suggestion to keep safety committee meetings manageable: Besides the occasion when a particular person cannot attend because of a more pressing responsibility, the number of members can be reduced by the general superintendent who can represent all his or her superintendents for a particular meeting and brief them later. Similarly, the union representatives can reduce their numbers if there are no pressing subjects to be discussed by a particular craft provided that the communication is sufficient between them and that updating and briefing follow at a later date for all those missing. Good communication is paramount both in and outside the meetings.

Reporting System

Much of any discipline of work these days requires reports; this is true of the safety activities as well.

It is very important to realize how helpful good documentation really is. It can substantiate any number of disputed facts. It encourages attention to detail, and that, together with good planning, is the secret to the success of any activity, particularly good site safety.

The safety program can be very easily quantified into specific actions and documents (see Figure 1.10). These documents have been found to be the manageable minimum.

From these sample forms it can be seen how little work is really involved in completing them (see also Forms 1.1 through 7 at the end of this chapter). Further elaboration of the documents is shown on the same brief schedule (Figure 1.10) with suggestions for distribution associated with the committee's reporting functions. Also detailed is a suggested basic filing system (Figure 1.11) to help the person who is familiar with safety but who may be inexperienced at setting up and coping with the necessary administration systems. All the forms shown in this chapter are just examples—they do not have to be used exactly as they appear here and they should be modified as required.

Frequency/Item	Form Used and Comments	Distribution
As Needed		
1. Employee safety indoctrination	Form 1.1	Copy to JF
2. Worker's compensation first report	Insurance company form of employer's report of occupational injury or disease	Original to IC, copies to IAC, LP, CCD, JF
3. Accident investigation report	Form 1.3	Copies to CCD, JF, ECM
4. Accident register (for every accident regardless of severity)	Simple log book	Original at job site, copy to CCD
5. Return-to-work slip	Self-explanatory	Original to IAC, copies to JF, LP
6. Near-accident report and evaluation	Self-explanatory	Copy to JF
7. Safety deficiency report	Form 1.7	Copy to JF, ESS
8. Crane yearly inspection and load test certificate	Supplied with crane by owner	Copy to JF
9. Equipment and vessel certification	Supplied by manufacturer	Copy to JF
10. Hazardous chemical material safety data sheets	Supplied by manufacturer	Copy to JF
11. Advertising, identification, information	Supplied by corporate company and federal government	Copy to JF
12. Test and approval certification for all first aid and safety equipment	Supplied by manufacturer and medical advisor	Copy to JF
Daily		
1. Daily job-site inspection	Small notebook	To JF on completion
2. Crane inspection	Form 1.6	Copy to JF
Weekly		
1. Toolbox meeting	Form 1.2	Copy to JF
2. Site fire extinguisher inspection	Form 1.5	Copy to JF
3. Safety committee progress meeting including subcontractors	Minutes of meeting	Copies to JF, AA, ECM, ESS
4. Weekly job-site inspection	Small notebook and report	To JF on completion
Monthly		
1. Crane inspection	Supplier provides forms required	Copy to JF
2. Job-site inspection	Form 1.4	Copies to ESS, CCD, JF, SM
3. Safety committee inspection and meeting	Minutes of meeting	Copies to ECM, JF
4. OSHA 200 Log	OSHA form	Copies to CCD, JF

Note: All originals go to the corporate loss-prevention and/or safety department unless stated otherwise. The abbreviations used in the distribution column are defined as follows: JF, job file; CCD, corporate construction department; ECM, each committee member; LP, loss-prevention department; AA, all attendees; ESS, each site supervisor; IAC, insurance administration corporation; IC, insurance carrier; and SM, site manager.

Figure 1.10 Safety program paperwork requirement schedule.

1. Orientation forms with each employee's signature and one complete set of orientation information.

2. Accident reports

3. Accident register log

4. Accident investigation reports

5. Workers' compensation/accident report and associated correspondence

6. Return-to-work reports

7. OSHA 200 log reports

8. Near-accident reports and evaluations

9. Toolbox meeting reports and worker complaints and suggestions

10. Subcontractor safety agreements and orientations

11. Fire extinguisher inspections

12. Safety committee meeting minutes

13. Monthly site safety and safety equipment inspections

14. First aid, inventory approval certifications, and orders

15. Industrial hygiene information and material safety data sheets

16. First aid, CPR, and safety course training and site personnel records

17. Equipment and vessel test certificates

18. Crane inspections, crane owner's certification

19. Company safety information

20. Fatalgrams and other important official distributable information

21. Equipment damage and insurance reports

22. Safety committee observation reports

23. Correspondence, segregated as required

Figure 1.11 Suggested filing system.

Some of the information may be superfluous to your particular project. If that is the case, delete the offending items. There are no hard and fast formats for reporting, as long as the information contained is understandable and the intent is fulfilled.

Sample Forms

The remainder of this chapter consists of examples of the following forms: Form 1.1 Employee Safety Orientation; Form 1.2 Weekly Toolbox Meetings; Form 1.3 Supervisor's Report of Accidents and Investigations; Form 1.4 Monthly Construction Inspection Checklist; Form 1.5 Fire Extinguisher Inspection Report; Form 1.6 Crane Operator's Daily Inspection Report; Form 1.7 Safety Deficiency Report; Form 1.8 Activity Hazard Analysis Form; and Form 1.9 Activity Hazard Analysis Worksheet.

Form 1.1 Employee Safety Orientation

I have received and read (or had read to me) the following items which are marked:

General construction safety requirements   [ ]

Corporate statement of safety commitment   [ ]

Craft safety practices (list each trade manual)   [ ]

_____   _____

_____   _____

Project rules and regulations   [ ]

Emergency evacuation procedures   [ ]

Emergency treatment procedures   [ ]

Fire control procedures   [ ]

Job-site safety program   [ ]

Others (list):   [ ]

_____   _____

_____   _____

I agree to follow all the instructions and procedures contained in this orientation.

I (have [ ] have not [ ]) successfully completed a Red Cross first aid course.

I (have [ ] have not [ ]) successfully completed a CPR course.

I completed these courses on _____.
                                                    Date

The Occupational Safety and Health Act of 1970 requires me as an employee to cooperate in complying with the act under direction of company representatives.

                              Signed _____
                                                    Employee

Safety officer or designee _____   Date _____

Form 1.2 Weekly Toolbox Meetings

Meeting Date: _____

Business Discussions

Subjects discussed: _____

_____

_____

_____

_____

Suggestions: _____

_____

_____

_____

_____

Actions to be taken: _____

_____

_____

_____

Injuries and/or accidents reviewed: _____

_____

_____

_____

Employees present: (list those present on reverse side)

Number in crew: _____ Number attending: _____

Craft safety representative: _____

Form 1.3 Supervisor's Report of Accidents and Investigations

It takes longer to report an accident than to prevent one.

Project: _____ Location: _____

Name of injured: Age: Length of service: Occupation:

_____   ____   _____   _____

Date of accident:_____ Time:_____

Work location at time of accident: _____

Extent of injuries: _____

Describe accident (state what injured was doing and circumstances leading up to accident):

_____
_____
_____
_____
_____

Check one in each section:

Accident type:

[ ] Fall, at same level
[ ] Fall, at lower level
[ ] Struck by
[ ] Struck against
[ ] Handling material
[ ] Caught between
[ ] Operating equipment
[ ] _____
[ ] _____

Unsafe practice:

[ ] Unsafe equipment
[ ] Unsafe loading, piling
[ ] Unsafe position
[ ] Working on moving machinery
[ ] Using without authority
[ ] Using at unsafe speed
[ ] Not using protective equipment

Unsafe condition:

[ ] Not guarded
[ ] Defective
[ ] Poor housekeeping
[ ] Safety equipment
     not used or
     provided
[ ] Poor lighting

Investigation:

Investigation made
   at scene_____ Yes
           _____ No
Discussed with
     (injured)(witness)
     (other) _____ Yes
           _____ No

[ ] Lost time
[ ] Serious injury*
[ ] Doctor case
[ ] First aid
[ ] Near accident

* Serious injury: (a) any work restriction or a physician's prescription for light work as a result of work injury; (b) laceration requiring mechanical closure (sutures or slips); (c) fracture; (d) work-caused eye injury treated by physician; or (e) injury causing lost time.

Why in your opinion did unsafe practice occur (worker lacked skill, misunderstood, lacked instruction, etc.)?

_____
_____
_____
_____

Why in your opinion did unsafe condition exist (hidden defect, not recognized, no authority to correct, etc.)?

_____
_____
_____
_____

Form 1.3 (Continued)

What action has been taken to prevent recurrence?

_____

_____

_____

If action not taken, why not?_____

Witness:

_____

Supervisor: _____   Manager: _____

Date of this report: _____

The unsafe acts and conditions that cause accidents can be corrected only when they are *known*. It is the *supervisor's* responsibility to *find* them, *report* them, and *state* their *remedy*.

Use the reverse side for additional comments and sketches.

All questions must be answered and the report sent to the corporate office responsible within 48 hours.

All property, tools, or material damaged as a result of this accident should be reported on the *First Report of Loss* form.

Form 1.4   Monthly Construction Inspection Checklist

Date: _____ Location: _____

Subject	Yes	No	Remarks
1. GENERAL (Conditions of Job Site)			
a. Posting OSHA and other job-site warning posters			
b. Records of safety meetings up to date			
c. Toolbox meeting minutes records up to date			
d. Availability of first aid equipment and supplies			
e. Job-site OSHA documentation up to date			
f. Emergency telephone numbers such as police department, fire department, doctor, hospital, and ambulance posted in the proper places			
g. General cleanliness of working areas			
h. Regular disposal of waste and trash in the proper manner			
j. Passageways and walkways free from obstruction			
k. Adequate lighting of all work areas			
l. Nails removed or bent over on all lumber and fixtures			

Form 1.4 (Continued)

Subject	Yes	No	Remarks
m. Proper chemical waste containers used for disposing of segregated waste in the appropriate manner			
n. Sanitary facilities adequate and clean; more than one unit for 20 persons			
o. Good supply of clean drinking water			
p. Sanitary drinking cups available			
q. Correct head protection used by all employees outside offices			
r. Eye protection readily available			
s. Face shields readily available			
t. Respirators and masks readily available			
2. ELECTRICAL INSTALLATION a. Adequate wiring, well insulated			
b. Fuses provided			
c. Ground fault circuit interrupters in place			
d. Electrical dangers posted			
e. Proper fire extinguishers provided			
f. Terminal boxes equipped with required covers			
3. HAZARDOUS CHEMICALS a. Fire hazards identified			
b. Proper types and number of extinguishers easily accessible			
c. All containers clearly, indelibly identified			
d. Storage practices compatible with safety regulations			
4. WELDING AND CUTTING a. Screens and shields used for personal protection			
b. Protective clothing available and used			
c. All equipment in safe operating condition			
d. Power cables undamaged and protected			
e. Fire extinguishers of proper type readily available			
f. Regular inspection for fire hazards			
g. Flammable materials well ventilated and well away from hazards			
h. Gas cylinders secured upright			
i. Gas lines in good condition and protected from hazards			
j. Cylinder caps used in the correct manner			

Form 1.4 (Continued)

Subject	Yes	No	Remarks
5. TOOLS (Hand Tools) a. Employee-owned tools checked on and off site and inspected			
b. Damaged tools repaired or replaced promptly			
c. Proper tool being used for the job			
d. Safe storage, adequate facility for carrying safely			
TOOLS (Power Tools) a. Proper grounding used for each set of tools			
b. Proper training carried out			
c. All mechanical safeguards in working order			
d. Right tool being used for the job			
e. Tools neatly stored when not in use			
f. Tools and cords in good condition			
g. Good housekeeping for tools and auxiliaries			
6. MACHINES AND EQUIPMENT a. Safety goggles or face shields used where required			
b. Tools used where recommended			
c. Proper training for machines and equipment carried out			
d. Compliance with all local laws and ordinances			
e. All operators qualified to operate respective machines			
f. Machines in good working order			
g. Machines and equipment protected from unauthorized use.			
7. GARAGES AND REPAIR SHOPS a. No fire hazards			
b. Fuels and lubricants dispensed cleanly and safely			
c. Good housekeeping throughout shop			
d. Lighting adequate for safe working			
e. Carbon monoxide dangers identified			
f. All fuels and lubricants stored in proper containers in a properly ventilated area			
8. FIRE PREVENTION a. Phone numbers of fire department adequately displayed			
b. Records of fire instructions, planning, and training given to employees			

Form 1.4 (Continued)

Subject	Yes	No	Remarks
c. Fire extinguishers checked and documented			
d. "No smoking" displayed where necessary			
e. Fire hydrant access to public thoroughfare open			
9. HEAVY EQUIPMENT a. All lights, brakes, warning signals operative			
b. Regular inspection and maintenance shown			
c. Dates of lubrication and repair of moving parts recorded			
d. Roadways well maintained, accessible, and clear of interferences			
e. When equipment is not in use, it is unable to be used by unauthorized personnel			
f. Wheels chocked when necessary			
10. MOTOR VEHICLES a. Local and state vehicle laws and regulations observed			
b. Qualified operators			
c. Regular inspection and maintenance shown			
d. All brakes, lights, and warning devices operative			
e. All glass in good condition			
f. Weight limits and load sizes controlled			
g. Personnel carried in a safe manner			
h. Fire extinguishers installed where required			
i. Reversing alarms provided			
11. SCAFFOLDING a. Scaffold tied in to the structure correctly and safely			
b. Scaffold walkways free of debris, snow, ice, and grease			
c. All workers adjacent to scaffold protected from any falling objects			
d. All connections checked for rigidity			
e. Guardrails, intermediate rails, and toeboards erected and sufficient			
f. Erected in accordance with the regulations and checked			
12. LADDERS a. Properly secured to prevent slipping or falling			
b. Ladders inspected regularly and in good condition			

Subject	Yes	No	Remarks
c. Handrails extend 36 inches above top of each landing			
d. Metal ladders not used around electrical hazards			
e. Ladders not painted			
f. Safety feet in use			
g. Stepladders fully open when in use			
h. Rungs or cleats not over 12 inches on center			
i. Job-built ladders constructed of sound materials and meet regulations			
13. BARRICADES a. Floor openings planked over or adequately barricaded			
b. Adequate lighting provided in all work areas			
c. Traffic controlled to all work areas			
14. MATERIAL HANDLING AND STORAGE a. Materials stored or stacked safely			
b. Fire protection available			
c. Stacks on firm footings not stacked too high			
d. Clear passageways around material			
e. Proper number of workers for the working conditions			
15. EXCAVATION AND SHORING a. Excavations barricaded properly			
b. Correct shoring used for soil and depth			
c. Adjacent structures properly shored			
d. Equipment at a safe distance from edge of excavation			
e. Proper ladders provided where needed			
16. STEEL ERECTION a. Hard hats, safety shoes, and gloves being used			
b. Taglines and safety belts being used			
c. Floor openings covered and barricaded adequately			
d. Ladders, stairs, or other safe access provided			
e. Hoisting apparatus checked regularly			
f. Safety nets or planked floors being used			
17. HOISTS, CRANES, AND DERRICKS a. Outriggers used when required			
b. Cables and sheaves inspected regularly			

Subject	Yes	No	Remarks
c. Loading capacity at lifting radius below the limit for the unit			
d. Signalers used when necessary			
e. Equipment is firmly supported when lifting			
f. Signals understood by operator			
g. Power lines inactivated, removed, or at a safe distance from all lifting exercises			
h. All equipment properly lubricated and maintained			
i. Inspections and maintenance logs kept up to date			
18. CORPORATE ADMINISTRATION a. General policy statement by senior corporate management with regard to safety and hazard control posted on site			
b. Policy (1) incorporated in employee indoctrination and (2) posted in a conspicuous place			
c. Conspicuous official notice boards on site			
d. Written safety regulations exist (1) on site,(2) are incorporated in orientations,and (3) are explained to all employees			

Identify type and location of unsafe acts and/or practices observed:

_____

_____

_____

_____

Site manager: _____

Site safety coordinator: _____

Form 1.5 Fire Extinguisher Inspection Report

Date: _____

Serial Number	Type	Size	Location and Condition	Date of Last Charge

Form 1.6 Crane Operator's Daily Inspection Report

Make: _____ Model: _____

Daily Visual Checks   *Initial each item as checked*

ITEM TO BE CHECKED	DATE					
1. Broken or cracked glass						
2. Damaged or missing guards or gear or chain case covers						
3. Drive chains and sprockets for cracked or broken pieces						
4. Oil or coolant leaking below rotating bed or car body						
5. Roller path, house rollers, and hook rollers for chips or cracks						
6. Boom hoist, whip line and hoist wire rope; pendants; load blocks; and sheaves						
7. Fuel tank(s); fuel gauges; and hose and connections						
8. Limit devices; boom/mast stops; drum pawls						
9. Control valves; levers and linkage; and instrument panel(s)						
10. Fire extinguisher available and in working order						

Daily Preventive Maintenance Checklist   *Initial each item as checked*

ITEM TO BE CHECKED	PROCEDURE	DATE					
Radiator coolant	Check level and add when necessary						
Hydraulic system(s) level	Check reservoir, and add oil when necessary						
Gear case lube	Check level, and add when necessary						
Engine oil							
Transmission and/or chain case or reservoir							
Rotating bed sump (if applicable)	Check level, and add oil when necessary						
Converter input and/or output housing(s)							
Air compressor							

Remarks:

Date: Item: Date corrected:

  Initials

Form 1.7 Safety Deficiency Report

```
[  ] VIOLATION OF RULES          Date: _____
[  ] UNSAFE ACTION               [  ] UNSAFE CONDITION

Observed: _____
_____
_____
_____

Where observed: _____
_____
_____

Action taken or recommended: _____
_____
_____
_____

Observed by: _____
Project manager/superintendent review or action: _____
_____
_____
_____

Project manager/superintendent: _____  Date: _____

Distribution:  Original: Supervisor of area of employees concerned
               Pink: Project manager/superintendent
               Yellow: Safety supervisor and job file
```

Form 1.8 Activity Hazard Analysis Form

```
Activity definition:

Activity location:

Simultaneous activities involved at the same location:

Crafts involved:

Assessed hazards (see worksheet):          Possible protections:

Relevant historical data and the unpredictable:




Practical protection recommendations:



```

Form 1.9 Activity Hazard Analysis Worksheet

The purpose of this sheet is to examine the following questions and establish the answers to them after evaluating the basic conditions and activities.

By answering these questions and others that may germinate from them, this complete analysis can now be evaluated and condensed to insert on the activity hazard analysis sheet.

1. Look at the critical-path activities on the project schedule; look for simultaneous phases which indicate simultaneous activities.

2. Define those simultaneous activities: where, when, and how do they occur?

3. Review these simultaneous activities to discern all the existing and predictable hazards that may occur. Look for those hazards that are not readily apparent.

4. Enlist head office and/or corporate or safety personnel to help define previous similar work and preventive actions; historical records of activities and accident frequencies; and types and severities of accidents.

5. Specify all the possible protections available.

6. Analyze and evaluate all the protections and enumerate all the practical protections.

7. Arrange necessary discussions, make necessary adjustments to the program, and prepare for execution of the preventive measures.

8. File and distribute the data for future use.

2

Elimination of Potential Safety Hazards

There are two major factors in a safety program: regular site inspections and detailed hazard analyses. These factors are most effective when used together, and their use should be the top priority of any safety program, no matter how small.

With the structure of the safety committee set up as described in Chapter 1, site inspections can now be carried out with some flexibility at several different levels.

Daily Job-Site Surveys

Job-site surveys should be carried out daily by everybody on the committee who works in the field independently of each other but confined to the respective areas of that person's activities. The craft representatives should be responsible only for their own craft, and should they notice anything outside their responsibility, they should report it to the appropriate supervisor for correction. This system thus delineates responsibility from support.

Daily site survey by each craft representative and subcontractor of his or her own areas

Craft representatives should recognize unsafe conditions and actions and rectify them, normally using their own work force.

For example: The carpenters are working on concrete forms. Their representative should regularly survey his or her area for hazardous debris, power tools improperly guarded, power cables improperly protected, and other potential hazards. Similarly, other trade representatives should do the same thing in their respective areas.

Any unsafe conditions noticed during these surveys should be rectified immediately using the trade representative's own work group. That way the problem stays within the trade and is resolved by people working within their normal expertise.

Daily site surveys also encourage communication among trades. Such communication is the single most important tool to be used in this approach to safety control.

Daily site inspection by management personnel

A very efficient and effective way of keeping on top of hazard recognition is by making it company policy that all management personnel are responsible for safety irrespective of whether or not they serve on the safety committee.

These responsible management delegates should be inspecting their areas concurrently with the trade surveys, paying particular attention to the "overlap areas." These overlap areas are probably the most critical to the program's success.

Overlap is defined in this context as those areas where two or more trades are working in restricted or confined spaces or are using the same areas for storage of materials and/or for access. An overlap area is where the most confusion occurs and is thus where the greatest potential for hazards and accidents exists.

Obviously, each trade is usually very capable of working safely in its own environment. Each crew is independently well coordinated. The problems occur, for example, when ironworkers are working above carpenters who are working by the side of laborers who are working close to the electricians who are laying underground conduit. These areas are where the management delegate should be concentrating. The constant changing of conditions and each craft's demand for space need constant attention.

Again, any hazards or potential hazards recognized during these inspections should be eliminated immediately.

Daily inspection by the safety coordinator

The final daily inspection routine should be carried out by the safety coordinator or his or her designee. This person will at least once a day, or more often with time permitting, inspect the whole site, paying attention to all site activities but particularly to potential hazards and hazard analysis, which will be elaborated on later in this chapter.

It is a good idea for each member of the committee to carry a small notebook to record unusual events, changes, and corrections. Such notes collected by the committee member are not for general publication but are for personal reference during safety and other coordination meetings. As time goes on, it is often surprising to the committee member how much he or she relies on the notebook.

Weekly Job-Site Inspections

The second phase of inspections is the regular weekly inspection of the whole site. This inspection is carried out, if possible, by two or three committee members, jointly with a minimum of one management delegate. The focus of this tour should be determined by the week's safety meeting highlights, the previous week's problems, and the future week's activities.

It is recommended that the committee rotate this group's members each week. That way, the group has a different outlook on each inspection, which should therefore reduce the possibility of missing something during the inspection. The group's observations should all be referred back to the safety coordinator for immediate action. Outstanding items would be tabled at the safety committee meetings. This practice encourages positive action on problems which occur and ensures that resolutions to outstanding items are not forgotten.

Monthly Job-Site Inspections

This third type of site inspection is very different from the other inspections, which are generally informal. Monthly site inspections are designed to keep on top of general site safety.

The monthly site inspection is a very detailed, formal inspection recorded on provided forms and performed by assigned personnel of the safety committee. (Different committee members are rotated each month, but as before, at least one person from site management goes on the inspection.) This inspection covers all aspects of the job site. (See Chapter 1, Form 1.4.) The inspection team prepares a written report which is then evaluated by the safety committee at their next meeting. The committee then distributes the results to all supervisors on site for resolution, rectification where necessary, and improvement by the responsible persons. Several days later the safety coordinator and the site manager meet to review progress and to implement any necessary further steps to ensure compliance with the safety program. The status and degree of acceptability is then reviewed in detail at the next regular safety committee meeting.

Not only do these regular inspections highlight problems, but they also highlight particular individuals and groups who do not have the same respect for the rules as the rest of the employees. Then, of course, once a problem is visible, it is very easy to correct by one means or another. Regular inspections also encourage extra care by everybody on the site. This step alone goes a long way to avoiding many of the silly accidents that occur through sloppy work practices and carelessness.

The second major thrust of the safety program is hazard analysis. There is nothing to be gained from sending somebody into the field to inspect a project for potential hazards if he or she does not know what they are or where to look for them.

Analyzing a task or series of tasks and assessing the hazards in that work are not easy. But most professional tradesmen are being trained by their union organizations to look for and understand these hazards.

Similarly, the management side is also training its people. Therefore, it is absolutely necessary to pick qualified-competent persons from both sides to work on the safety committee and to perform inspections and hazard analysis. Through regular communication and supplementary training, the work force then becomes more aware, more responsible, and more able to identify and eradicate a hazard.

Historical Data Hazard Analysis

The first section to start with is assembling historical records for the industry, the locale, the corporation, and each trade union involved. Then try to accumulate from these sources, plus, of course, OSHA, NIOSH, and your insurance company, all the typical accidents that have occurred with the construction and operations of this type of project. Such research can be a laborious task, but many organizations have this information readily available, so a series of telephone calls can very quickly produce an extensive list.

Many different organizations have developed safety statistics of various types, for example, the National Safety Council, the Bureau of Statistics, insurance companies, and the federal and state departments of labor. All this information is now accumulated and can be analyzed to ascertain the most frequent accidents by type and by craft as related to any particular project or plant. This is probably the most important information to develop. Then, armed with these facts, a detailed training session can be held to disseminate this information to each craft representative and superintendent on site. Having highlighted probable exposures, it is much easier to eradicate the opportunity for these exposures to exist. By word of mouth, through toolbox meetings, and by visual inspections, everybody is therefore more aware of what to look for and where to look for potential hazards and accidents to happen.

Some basic examples follow. These facts have been accumulated from a cross-section of states and therefore closely represent the national average.

Laborers' most common reasons for lost-time accidents are:

1. Overexertion

2. Being struck by an object

3. Being struck against an object

Ironworkers' most common reasons for lost time are:

1. Falls from elevations

2. Overexertion

3. Being struck by an object

From this basic information, it is very easy to ascertain that in the case of laborers, special attention should be paid to lifting properly; i.e., do not bend forward to lift, but bend the knees and keep the back straight; do not carry too much weight; and warm up the body before any undue effort, especially in cold weather.

Other basic assumptions can be made from the information associated with laborers and distributed to the work force.

With ironworkers, the highest exposure is falling from elevations. Therefore, particular attention should be paid to, for example:

1. When working above the ground, the craftspersons should be wearing safety belts and tying off properly.

2. Floor penetrations should be securely protected.

3. Building open-floor perimeters should be properly barricaded.

4. Ladders should be properly tied off, etc.

With these types of historical data studies, certain assumptions can be made and implemented, and exposure can be reduced. These studies do *not* take long—even the corporate safety department may be able to help in accumulating this information—and the results are very worthwhile.

Planning and Site Hazard Analysis

There are many positive results from using a safety committee effectively. One of the most important is eliminating exposures by forethought and planning. This whole subject will be dealt with in great detail in the next chapter, where it will be addressed in its relation to existing construction procedures. The purpose of this section is to explain that many steps may be taken without these construction procedures being used at all.

For example, large pieces of equipment are coming onto the site. Weeks prior to the arrival of the equipment, the safety committee starts checking weight restrictions, ease of lift, correct equipment for lift, size of slings, quantity of workers, assignment of responsibilities to those workers, etc.

Most accidents happen due to lack of concentration or forethought and/or disorganization. By planning, a great many accidents can be eliminated.

Another example: Hazardous chemicals are coming onto the site in 2 months' time. Using the material safety data sheet, the safety committee considers the following issues in their plans:

1. The controls and location for access and storage are safe.
2. The correct antidotes are available for personnel and environmental protection.
3. Protection, e.g., glasses, gloves, and breathing apparatus, are available.
4. Local hospitals are alerted that these chemicals will be used so they will be ready in the event of an accident.

Each activity in construction should be planned and analyzed for hazards. For example, several near-misses have occurred in using hand-operated grinders. The safety committee has analyzed the probable causes (this activity will be discussed in more detail in Chapter 6).

They have determined that further training in the use of this type of tool is required, and they have therefore recommended that the safety coordinator expedite a training course for site personnel. They also feel that the grinding wheels are not as good as they should be. The committee again recommends that the coordinator expedite immediate action by the supplier and perhaps restrict use of the grinders until the problem is resolved.

The result of the safety committee's action is that the opportunity for the accident to happen has been reduced. By eliminating the opportunities for an accident to happen and by eliminating as many of the conditions conducive to the accident as possible, *the accident will not happen.*

To draw a parallel: Three elements make up a fire: combustible material, heat, and oxygen. Remove any one of the three elements, and the fire will not happen. The same applies to accidents: Remove the cause or any of the circumstances which promote the accident, and it will not happen.

Planning takes away some of the elements conducive to an accident's occurring. There is an old maxim which is very relevant to these circumstances: "An ounce of prevention is worth a pound of cure."

Worker participation

Participation in the site safety activities should be encouraged in all site personnel, including the entire work force. Such participation will result in a safe clean site. Examples of this type of planned activity are:

1. Run a site competition for the best safety slogan. For example, the following is a past winner:

   The *Four Important G's* in safety: *gloves, guards, glasses,* and *government regulations.*

2. Encourage anybody seeing a hazard to identify that hazard to the safety committee. Keep records of the industrious individuals who bring such information to the attention of the committee, and reward them at the end of the month. It is of paramount importance to have this rapport and enthusiasm on site with the whole work force for the program to really work.

3. Run a site competition for the cleanest work area by the month. Prizes can be company sales material, e.g., a pen and pencil set or golf balls, or they may be even more ostentatious, e.g., negotiable air tickets. With a little bit of inventiveness, all sorts of encouragement may be designed to encourage a clean, safe job site with all the employes cognizant of hazard control.

The doubters will argue, "Corporate offices may not sanction this unnecessary additional expense." However, the actual expense is far less than may be expected. And all prize items are tax deductible! A clean, safe job site may increase productivity by 5 percent and has been responsible in one case for a 12 percent increase in productivity. Equally important are the following bonuses to the company:

1. Lost time caused by accidents is reduced drastically.
2. Insurance rates are reduced.
3. Job-site attitudes are better.
4. Company reputations improve.

All these tangible, quantifiable attributes and more occur. No, it is not a great expense!

Planning of purchasing and training information

A wise financier once said, "You have to speculate to accumulate." This principle applies to a well-run safety program as well as to high finance. Spend a little more on a good safety program and save money for the rest of the project.

Planning means having the right tools on site to do the job properly and safely, even if that means spending a little more on better quality to make it safer. Planning means using company and federally published safety information to its best advantage and making sure it is well distributed.

For example, the safety committee should distribute craft training cards—brief reminders of peculiar safety hazards in each trade—highlighted on a durable card which tradesmen can carry around in their top pockets (see examples in Figures 1.3 to 1.6).

Planning a clean job site

Finally, a brief note on safety planning: It has been found that many recent successful projects (even the small ones) have benefited supremely from the policy of using two laborers, under the direction of the labor foreman, purely on cleanup and safety-related matters for the whole job. The job stays cleaner, tidier, and safer, and there is a flexible work gang available for any safety-related activity as is necessary. This practice prevents a particular craft's being pressured into working in hazardous conditions because nobody is available instantly to do the preparation necessary for the job to begin properly. An old saying comes to mind: "Haste makes waste," meaning that tasks done in haste are finished more slowly and less safely. Another old saying is equally accurate: "A clean job site is a safe, healthy job site."

It is again interesting to note that using two laborers purely for cleanup and safety-related matters is actually not an added expenditure. It is an actual saving. For example, when VIPs come to the job site, there is usually a 2-day massive cleanup just prior to their visit, a practice that reflects a large production loss. An untidy work site slows down workers, which is also a large production loss. And an untidy work site allows more hazards to exist, which creates opportunities for lost-time accidents. All this also means larger insurance premiums. One or two laborers dedicated full time to cleaning up and to safety activities pays for itself many times over, even in the course of a small job.

The final, and perhaps the most important, new philosophy in construction safety is to attempt to use a policy introduced several years ago into industrial safety (machine shops and process industry, etc.). This policy has proven to be the tool that significantly alters and reduces accident trends and it is simply called "job hazard analysis."

Job Hazard Analysis

Job hazard analysis (JHA) in general industry is the "ultimate concept" in safety control and is going to be a major milestone to accomplish in any shape or form. And with all the talent in the construction industry and with both management and labor contributing, the end result can be accomplished. By way of explanation, in general industry the JHA system has been working for several years and has accomplished excellent results. JHA is a simple program, which basically takes any particular craft action in industry, for example, operating a drill press, and analyzes each action (as in a time-and-motion analysis) to evaluate the safe, efficient way to perform that action or series of actions. From this point it is relatively easy to train the operator to perform the action in a prescribed manner. Thus a JHA will eliminate many of the opportunities for an accident to happen during the performance of a particular task.

With construction activities, however, a JHA is extremely difficult because each action occurs in so many different environments and conditions. But the industry has to attempt to categorize these activities and start applying the same logic.

JHAs can be utilized in construction, and this is one method: Lay down broad guidelines to be able to analyze any required activity, or group of activities.

1. Define the activity and the trades involved.

2. Define the actions involved in this activity and the hazards.

3. Define the tools, equipment, and materials to be used in conjunction with this activity.

4. Define the various places on the site where this activity will be performed in the module of construction being analyzed.

5. Define the other work adjacent to this activity being performed concurrently which could cause interactions or interferences and analyze any additional hazards.

6. Define the protections and procedures that can be used to eliminate both the existing and the predictable hazards.

These steps can be subdivided, but to do so would make the system very cumbersome and lengthy,* which is counterproductive.

---

* It is important to realize at this point that tradesmen are trained thoroughly in their fields of expertise prior to performing any craft in the field. Therefore, the individual task is *not* the basis for analysis. The *group* of tasks in any series of activities is the basis for analysis, and how they interfere and interact with each other.

From this information which will have been tabulated as shown in Figures 2.1 and 2.2, the evaluation will take place to find the most efficient manner in which to perform the activities and with what forms of protection.

At this point, in order to avoid confusion, it is probably a good idea to provide some definitions of prominent words used in this discussion.

Activities

The term "activities" refers to a series of tasks amalgamated to produce an end result and therefore an activity, e.g., steel foundation installation, concrete pouring, steel erection, and machinery installation.

Task

A "task" is a unit of work such as welding or drilling. Many tasks make up a complete activity.

Module

Each module is a group of activities which together complete a scheduled part of the project. A project is split up into several modules, e.g., excavation, foundations, concrete work, and structural erection. These modules are subdivisions of the four main phases mentioned in the preceding chapter (civil, structural, mechanical, and finishing).

In normal, but particularly in fast-track, construction the activities consist of performing tasks simultaneously in several concentrated areas, where other related or unrelated activities are also being performed.

For example, assembling and setting concrete forms activities are occurring adjacent to steel erection and machinery erection activities. Therefore, the activity hazard analysis should include all the tasks in the concrete formwork and include all the tasks in the steel erection and machinery erection as well. These are adjacent and/or interacting and/or interfering with each other. The existing and predictable hazards have to be identified and eliminated. Planning will also be used to predict the possible hazards and to make practical allowances for them, where feasible.

Form 1.9 (Chapter 1) lists the questions which should be answered by all the participants involved in the activity hazard analysis.

These answers should be detailed accurately on the activity hazard (Form 1.8, Chapter 1) analysis (AHA) as shown in Figure 2.1. This is where all the information is assessed and where the analysis and recommendations are completed. Finally, the document is distributed to all interested parties.

Activity definition:

Assembling machinery on its foundations

Activity location:

Column row 4 through column row 9
Elevation: Ground level up to elevation +120 feet

Simultaneous activities involved at the same location:

Steel erection, piping erection, conduit installation

Crafts involved:

Ironworkers, operating engineers, millwrights, pipe fitters, electricians, and laborers

Assessed hazards (see worksheet):	Possible protections:
Overhead cranes working	*Stop work beneath each crane lift*
Welding hazards: sparks and flashes	*Erect protective barriers with canvas*
Falling materials and tools	*Safety glasses with side shields*
Personnel falls	*Stagger work activities*
Area saturation (craftspersons)	*Safety belts*
Material and equipment saturation (leads, cables, hose, etc.)	*Arrange suspended cable routing*
Burns from welding, cutting, and burning activities	*Flag hot areas (mark or cover). Additional clothing protection (gloves, etc.)*
Poor lighting	*Increase lighting*
Cold area (no heating, conducive to frost and icing, poor foothold, slipping, etc.)	*Add temporary heating. Constant area inspections (by safety and craft supervisors)*
Piping (pressure testing, etc.)	*Remove personnel from test areas*
Space restrictions for peripheral activities	*Prefabricate well away from restricted area*
Fires	*Increase fire protection*
Working in confined spaces	*Additional craft-training-awareness sessions required*

Relevant historical data and the unpredictable:

Accidents in similar circumstances, falling objects (near-misses), burns, welder's flash, eye injuries (falling dirt and dust), two overhead crane accidents, three minor fires

Constant repetition of work through crowd movement and oversaturation

Previously used shiftwork for specific conflicting crafts (very successful)

Daily progress meetings

Practical protection recommendations:

Use shiftwork to alleviate specific craft interferences.

Increase safety inspections in the area.

Increase temporary lighting and heating.

Prefabricate well away from this area.

Safety belts and glasses with side shields are mandatory.

Prepare daily planning and progress meetings with craft supervision.

Erect temporary barricades (fireproof).

Increase available fire extinguishers and firewatch.

Prepare overhead crane schedule.

Figure 2.1 Completed activity hazard analysis form.

The purpose of this sheet is to examine the following questions and establish the answers to them after evaluating the basic conditions and activities.

By answering these questions and others that may germinate from these, this complete analysis can now be evaluated and condensed to insert on the activity hazard analysis sheet.

1. Look at the critical-path activities on the project schedule. Look for simultaneous phases which result in simultaneous activities.

2. Define those simultaneous activities: where, when, and how will they occur?

3. Intermesh these activities and discern all the existing and potential predictable hazards in the worst case. Even unpredictable hazards should be part of that process and defined.

4. Enlist head office and/or corporate or safety personnel to help define previous similar work and preventive actions. Obtain historical records of activities and accident frequencies, types, and severities.

5. Specify all the possible protections available.

6. Analyze and evaluate all protections, and enumerate all practical protections.

7. Arrange necessary discussions, make necessary adjustments to the program, and prepare for the execution of preventive measures.

8. File and distribute the data for future use.

Figure 2.2 Completed activity hazard analysis worksheet.

As these AHA documents accumulate, you will find that everybody's awareness of hazards increases, and at this point, it is hoped, the significant reduction in accidents or near-misses also become noticeable. It is absolutely amazing that when companies conduct both regular inspections and activity hazard analyses, the accident and near-miss incidence rate drastically drops. It is not magic—it is purely and simply good planning, good organization, and good communication.

Planning and Scheduling for a Safety Program

Planning is a critical area in the control and enforcement of a safety program. In these days of very tight, competitive bidding with the necessity to cut costs, there still is and must be a place in the budget for safety equipment and a safety program. A safety program is not a luxury. More and more organizations realize that not only is such a program mandatory but it is in fact a necessity in achieving efficient, safe sites which encourage repeat clients. Clients in all walks of trade and industry are being more and more vociferous in the demand for safer construction sites, and they are very supportive of good safety programs because they know a safe site is an efficient site and is therefore going to give them a faster, better, safer job.

The key to success for any project of any size and type is planning and scheduling.

Planning a Safety Program

Bid-stage planning

Bidding is the primary stage of planning, and at this point a realistic figure can be put on the cost of the safety program for any particular project. Industrial and commercial construction can be treated in the same manner for this purpose.

It has been seen over the last several years that indirect costs on any type of construction project are increasing; however, they are still less than 40 percent of the total project cost.

A safety program, as defined here, will cost less than 0.125 percent of those indirect costs of the project and include all the following items:

1. First aid boxes

2. Literature, warning signs, forms, etc.

3. Hazard-detection equipment, safety equipment

4. Fire extinguishers

5. Filing cabinets, office files, etc.

6. Training in hazard control

7. Videos and other materials for indoctrination

8. First aid and CPR training

9. Safety work group

These items together translate into costs of approximately 0.05 percent of the total project costs for a rewarding safety program.

Now let's add up a few other figures. Workers' compensation costs are computed as being an average 5 percent of the total labor cost on any job. These are predictable costs. General industry (not the construction industry) uses programs similar to the one explained in this book but obviously on a more sophisticated and complex level because they have been making a concerted effort in this direction for a longer period of time. They are now avoiding approximately 80 percent of what were predictable accidents in the course of a year by dedication to their safety programs. Therefore, the average of 5 percent of total labor costs on any job for workers' compensation is now reduced to 1 percent. Now wouldn't that be a significant savings on its own if the construction industry were to achieve the same thing? It is not that difficult. Planning and implementing a safety program will yield similar results in construction.

Now let's examine the proposal further. Highlighted and noted in the prebid (technical) documentation should be all the prevailing conditions and special equipment planned for use, for example, heavy machinery, heavy-lifting equipment, hazardous chemicals, and the existing job-site environment, together with the notes on any peculiarities of the prospective client's safety programs. All these should be available for future reference by the site staff. Proposed project schedules and personnel calculations are also developed at this time.

All this information is generally needed to put the bid together whether there is a safety program specifically included or not. So to keep the information and pass it on to the site when the proposal is successful and the job is awarded is no extra effort at all.

To many companies, bidding is so competitive now that even greater detail is paid to achieving the lowest price. Many use really inventive procedures to lower labor rates to achieve project completion. Safety is, or should be, always considered in this context by the estimators. Why not pass this information onto the site staff when the proposal becomes an actual project? You can see, I am sure, that armed with this information, the individual assigned to job-site safety now has saved several hours of his or her time. He or she already has the basis of a safety plan without having to lift a finger. It is all there in the proposal documents. Now all this information is going to be used effectively instead of just thrown into a corporate file and never looked at again. The link between the bid and the successful beginning of the project is established; now the stage is set, and the site work begins.

Planning for Mobilization and for the Subcontractors' Arriving on Site

The project superintendent–site manager and the site engineer have been selected and delegated to site operations. Primed with the necessary paperwork, forms, and equipment for safety-related matters, including first aid boxes, etc., they arrive on site. The safety program planning and scheduling starts.

1. The first duty is to meet the client and ascertain and implement his or her special requirements.
2. The second duty is to review the project master schedule, including the manpower curves. Develop, using realistic milestones, all required crew sizes and the program of work. Develop the project philosophy with regard to required subcontractors.

The following paragraphs contain some suggestions on working with subcontractors and their interpretations of a safety program.

Subcontractors

It is a fact of life that many subcontractors have one and only one object in mind when they arrive on site: They want to do their job with as few people as possible and in as short a time as possible—and everybody else had better stay out of their way.

Of course, this philosophy immediately creates problems for the safety-conscious, caring, and concerned general contractor. Some subcontractors are a pure pleasure to deal with and some are not. Once they know the rules, some work hard at them and some do not. Solutions are never simple, but the following suggestions are valid and do work, and they do help control the "wild and woolly" difficult subcontractor.

Ensure that there is a safety clause with financial implications relating to the general contractor in the subcontract bid documents and in his or her contract. Something like the paragraphs, listed in the next section should be in the general terms and conditions section of the contract.

Protection of work and safety

1. Subcontractor shall continuously maintain adequate protection of all its Work from damage and theft and shall protect Owner's and Contractor's property from injury or loss arising in connection with Subcontractor's performance of the Work.

2. Subcontractor shall take all necessary precautions for the safety of employees, visitors, invitees, and licensees on the Site and shall comply with all applicable provisions of federal, state, and municipal safety laws and regulations and building codes and with all instructions and regulations of Contractor's and Owner's safety departments to prevent accidents or injuries to persons on, about, or adjacent to any area where the Work is being performed. Subcontractor shall erect and properly maintain, at all times, as required by the conditions and progress of the Work, all necessary safeguards for the protection of workers and the public and shall post danger signs warning against the hazards created by its Work.

Subcontractor shall designate a responsible member of its organization, whose duty shall be the prevention of accidents. The name and position of the person so designated shall be reported to Contractor.

3. Subcontractor shall be responsible for any fines, penalties, or assessments levied against or imposed upon itself, Contractor, or Owner arising from or related to any failure or alleged failure by Subcontractor or the Work to comply with applicable health, safety, or environmental laws, rules, regulations, or standards.

Similarly, something like the following paragraph should be incorporated in the special conditions of the contract which would point out conditions peculiar to that particular project and job site:

Safety: Hard hats and safety glasses must be worn on site at all times. Failure to do so will result in a violation of the contract and will be penalized.

The General Contractor's job rules and safety regulations, principles of construction safety, and craft training cards are made part of the subcontract, and copies of these are attached hereto. Additional copies for your work force are available on site.

These two documents, general and special terms and conditions, are the basis, together with fair, firm, and equitable administration of the contract, of the rapport you develop with your subcontractor. These documents will avoid the age-old problems from becoming major stumbling blocks and encourage safe working conditions with subcontractors and their work forces.

It is also of paramount importance to ensure that at the prebid meetings, prejob conferences, and subsequent site-orientation procedures that the subcontractor has explained to him or her in very clear, concise terms the safety program and what is expected from his or her organization and work force. Document these meetings and get the subcontractor's representative to sign that document.

Now you have a basis to enforce the rules fairly but strictly to the letter without any complaints from the subcontractor of unfair or biased treatment. Make sure, however, that all the documents show consistent facts. There must be no opportunity for confusion or misinterpretation.

If these meetings are handled as they should be and treated with the degree of importance they deserve, most subcontractors will realize the implications and happily "toe the line." For the ones that do not, ensure that the first time there is a problem that action is taken. This will set the tone for all further disputes. Needless to say, weakness or indecision will be harmful to everybody and usually will cause a poor relationship from which there is no escape. Firm, fair action, on the other hand, usually returns the relationship to some form of sensibility and mutual respect.

A final word of advice: When subcontractors become difficult, use your labor relations department to support you—that is their expertise. Now let us get back to the project plan related to the safety program.

We have ascertained the manpower needed, number of trades, crew sizes, project length, subcontractors, and major items of equipment and machinery required. Now is the time to start the corporate safety office on the research requirements. What are the types of accidents most prevalent to this type of construction? What are the most frequent types of accidents for tradespeople in this area? How do the union organizations in the area train their workers in safety control and hazard identification?

Once you have the answers to these questions and the many others which will come racing to your mind, you can start planning how to train your employees, what to train them in, and the types of manuals, toolbox training information, material safety data sheets, safety emergency plans, and electrical grounding systems required. All of a sudden the safety program is starting to take shape.

The orientation information is now available, and construction, usually ground clearing and excavation, starts. The filing systems are all ready to be filled with reports and notes, etc.

Project Construction

The first site management group has developed the critical-path schedule and the manpower curves and major milestones; now is the time to plan the size and makeup of the safety committee. The rotations of trade representatives on the committee can be scheduled, and actions and recommendations can be carefully reviewed and itemized in terms of health, hygiene, and safety:

1. Major lifts and types of lifts, sizes of cranes, quantity of lifting tools, ropes, slings, manpower, and barricades, etc.

2. Trenching requirements for foundations and conduit (always a highly hazardous operation if not controlled properly), the bracing and sheeting required, tools, hazard-warning lamps and barricades, etc.

3. Hazardous chemicals to be introduced to the site either for use during construction or as part of the finished process. Documentation required, material data sheets, location for storage, controlled access, antidotes readily available, chemical disposal, etc.

4. Construction of high structures, specialized roofing, electrical requirements, existing plant hazards and power lines, temporary power, site facilities, portable toilets, etc.

All these items need careful attention in the site indoctrination procedure.

There are many steps the work force can take which can help with these assessments once they are an integral part of the safety program. For example, most trade organizations now run of their own volition:

1. Safety and health training programs

2. First aid and CPR training

3. Craft specialist training courses

These and other training opportunities are designed to support site management in the construction skills required to complete the project on or ahead of schedule by using very experienced craftspersons to teach the "tricks of the trade" to all the less experienced personnel within their specific trade.

Remember all craftspersons are professionals. A major part of their ability to be professionals relies on their skill to carry out their trade safely and efficiently. Any craftsperson who is considered a real professional must have the same high regard for others' welfare as for his or her own while carrying out his or her duties.

There are many professionals on a job site. Use them to help plan and support your project safety campaign. Their input can be very significant if the opportunity is given to them.

As the construction program begins to take shape, 14- or 21-day work schedules are developed; then 90-day schedules are developed with more finite planning, in order to assure that work progresses in accordance with the program.

During its formation, all these tools are used by the safety committee to evaluate site conditions, requirements, and the equipment and personnel needed to do the job safely and efficiently.

When shiftwork starts, if indeed it is required, make the necessary arrangements well in advance so that the changes occur as a smooth transition. *Each and every change in site conditions, site operations, and new construction is a potential hazard.*

Communication is of paramount importance; i.e., let as many people as possible know future plans and the direction of work. Planning and teamwork are of equal importance. This is easily achieved at the toolbox meetings.

The daily, weekly, and monthly site inspections take place, the planning becomes a reality, and the safety program is adopted as an integral part of the site construction program.

The whole work force participates, becomes interested and offers suggestions, and pools their ideas, and a safer, trouble-free workplace results. Usually the direct result is a successful, safe project. This can be attributed to planning, attention to detail, and communication.

Commissioning or Plant Start-up

Plant commissioning or start-up is a much underrated and underestimated activity which, if not properly addressed, really creates catastrophic problems and hazards.

This fact applies more to the industrial project rather than the commercial project, but it is worthy of note to any commercial contractor because there are cases where this is very relevant.

It is essential to have a company policy for safety tagging and machinery turnover (see Figure 3.1). It is also imperative to hold daily commissioning meetings so that all involved—department and craft supervisors—know what is going on where and when. A tagging system is mandatory; in many cases companies have been refused contracts because they did not have recognized tagging procedures. The suggested tagging system operated by the commissioning supervisor and supported by the safety committee works well, and this procedure does reduce the accident rate substantially for this phase of work. Accidents with moving machinery are usually severe, and they need to be avoided at all costs. Training sessions for all personnel by the commissioning supervisor and the safety committee should be held regularly and well in advance to ensure that everybody connected with the project, *including the client,* understands the tagging procedures clearly.

A key (or lockout) system usually goes hand in hand with this procedure and should be treated with great respect, to ensure that everybody obeys the rules.

Figure 3.1 Safety tagging system for commissioning.

1.0 PURPOSE
  To provide a uniform safety procedure for the specific purpose of protection of personnel from real or potential danger

2.0 SCOPE
  To cover safety tagging procedures during all phases of construction, preoperational testing, QA inspection, and plant commissioning. This procedure is to be used in conjunction with the owner's safety program.

3.0 SPECIAL REQUIREMENTS
  3.1 *Documentation Forms*

   3.1.1 Documents utilized in conjunction with the performance of this procedure include the following:

   Red tag: "Danger: Do not Operate."
   Yellow tag: "Caution: Testing in progress"
   Green tag: "This equipment has been turned over to the operating company."
   Red tag log: To control red tag placements and lock placements

4.0 PROCEDURE
  4.1 *Responsibilities*
   4.1.1 *Commissioning engineer.* Testing department will be responsible for administering this procedure during the construction phase and for tagging and administering this procedure for equipment and components in the preoperational test phase.

   4.1.2 The construction department will request tags for components still in the construction phase and request work clearance and tagging for rework or repair of components that are in the preop test phase.

   4.1.3 The supervisors of the personnel involved in work where a potential hazard exists are directly responsible for ensuring that the requirements of this procedure are followed.

   However, safety is everyone's responsibility, and all personnel on the job site must take an active interest in ensuring not only their own safety but the safety of their coworkers.

   4.1.4 The operating company is responsible for tagging systems and components that have been released for start-up, operations, and maintenance.

   4.1.5 It is the responsibility of *all* job-site personnel to abide by this procedure. Failure to do so shall be considered sufficient grounds for immediate dismissal.

  4.2 *Tagging*
   4.2.1 This tag will be used for isolation of equipment, i.e., circuit breakers, switches, valvues, etc., where a real or potential hazard would exist if that component were repositioned.

   4.2.2 The tag is requested by the personnel involved in the work or their supervisor and will be issued and placed by the testing department prior to commencing such work.

   4.2.3 The tag stub will be retained by the work personnel or their supervisor during the performance of the work.

   4.2.4 A separate tag must be placed for each crew involved in the work.

   4.2.5 While the tag is in place, the device to which it is attached cannot be repositioned under any circumstances.

   4.2.6 The tag may be removed only by the person holding the stub to this tag.

   Immediately following the completion of the work, the stub holder will remove the tag and surrender both the tag and the stub to the testing department.

   The testing department shall maintain a red tag log with entries for the issuance and clearance of all red tags.

Figure 3.1 (Continued)

When a tag is issued, the testing department shall enter the date, the tag and lock numbers, tag location(s), system, purposes, and the name of the stub holder.

When the tag is removed, the testing engineer that receives the tag, lock, and stub shall enter his or her initials and date on the red tag log.

The tag and tag stub shall be stapled together and maintained on file for the duration of the job.

4.2.7 *The yellow tag: "Caution—In test"*

This tag identifies the boundaries of the system and/or components under preoperational testing.

When equipment enters the preoperational testing phase, the testing department will hang yellow tags as follows:

Electric isolation breakers on switches.

Boundary valves or components.

Local and remote operating stations.

Motor(s).

Around the equipment generally so that the equipment is clearly identified as under test from every possible approach direction. The tags can be used in conjunction with barrier tape if considered necessary by the testing engineer.

When an item is yellow tagged, that component can be repositioned or operated only under the direction of the responsible testing engineer.

Personnel must exercise caution when working in the vicinity of yellow-tagged equipment. It must be assumed that this equipment could start at any time. Red tags must be requested if the operation of this equipment would pose a hazard to personnel working in the area.

No construction work or rework may be performed on a yellow-tagged system without proper work clearance from the testing department.

Yellow tags are removed at the time of turnover to the client when the green tags are placed.

4.2.8 *The green tag: "This equipment has been turned over to the operating company for start-up, operation, and maintenance."*

This tag designates the boundaries of the systems and components that are turned over to the owner for operation.

Only the owner or his or her designated representative may operate or work on a green-tagged system unless proper work clearance is authorized.

Green tags are placed at the time that equipment is turned over to the owner for operation and maintenance as follows:

Electric isolation breakers or switches.

Boundary valves or components.

Local and remote operating stations.

Motor(s).

Around the equipment or system generally so that the equipment or system is clearly identified as under the owner's control from every possible approach direction. The tags can be used in conjunction with barrier tape if considered necessary by the commissioning engineer.

Personnel must exercise caution when working in the vicinity of green-tagged equipment. It must be assumed that this equipment could start at any time.

Figure 3.1 (Continued)

Red tags must be requested if the operation of this equipment would pose a hazard to personnel working in the area.

Green tags can be removed by the owner only after the contruction effort is completed.

4.3 Locks and Lockout System

4.3.1 Locks shall be used in conjunction with the red, "do not operate" tags. Locks cannot be used in lieu of red tags and must have an accompanying tag when placed.

4.3.2 Locks will be serialized and will be issued by the commissioning engineer. The serial number of the lock will be noted in the red tag log.

4.3.3 Locks must be removed with the red tag as soon as the affected work is complete.

4.3.4 Locks shall not be used to maintain circuit breakers, disconnect switches, etc. in the closed position.

4.3.5 The key for the lock will be retained by the same person that holds the red tag stub.

4.3.6 A master key will be retained by the project superintendent and the commissioning engineer and will be used only in the case of a lost key as specified in Section 4.5 of this procedure.

4.4 Clearance

4.4.1 Work or rework required when system is in preoperational testing control.

If it is necessary to perform work on a system, work clearance will be obtained from one of the testing engineers. In this case the red danger, "do not operate" safety tags will be placed over the yellow tag, "caution—in test" tag, by a testing engineer. These tags will be placed on circuit breakers, valves, etc. to ensure a safe work boundary. These red tags will be placed and removed by a testing engineer. While work is being performed, the tag stubs will be in the custody of the supervisor in charge of the work.

The red tags cannot be removed until the work is complete and the supervisor surrenders the tag stubs to the responsible testing engineer.

If there is more than one type of craft involved in the work, then a separate tag will be issued for each craft and the stub will be retained by their respective supervisors.

4.4.2 Work or rework required on a system that has been turned over to the operating company.

If it is necessary to perform work on a system turned over to the operating company, the person requesting work clearance will fill out red safety tags and give them to the responsible testing engineer.

The testing department will contact the owner, and after the owner has isolated the equipment and tagged it out, then the red safety tag will be hung over the owner's tag, and the stubs from these tags will be given to the responsible supervisor.

When the work is complete, the supervisor will return the tag stubs to the testing department who will remove the red safety tags. The testing department will then contact the owner and inform him or her that the work is complete and the red safety tags are removed.

The owner will then remove his or her safety tags and return the system to a normal line-up.

4.5 Lost Safety Tag Stubs or Keys: If when the work has been completed and the safety tag stubs or keys have been lost or are not available, the responsible stub holder will send written notification to the commissioning engineer that the work is complete and

Figure 3.1   (Continued)

> the tagged system is clear. If the stub holder cannot be reached, his or her supervisor can issue this notification. After receipt of such written notification, the commissioning engineer, after inspecting the affected area, may authorize removal of the tags or locks.
>
> 5.0 PROCEDURE CLOSEOUT
>   5.1 *Red Tags:* The commissioning engineer will review the red tag log to ensure log closeout on a regular basis during the commissioning procedures.
>
>   5.2 *Yellow Tags:* The commissioning engineer will remove yellow tags from the equipment and components no longer under testing and control.

Demobilization

Finally in this section we deal with demobilization at the end of, it is hoped, a very successful project. This is another area that gets forgotten in the humdrum of a fast-track construction project: the pressure is off, and everybody relaxes. Well, not quite!

Attention should be paid at this time to the following areas:

1. All reusable safety equipment and tools should be inspected. Repair them as required or clearly mark them "needs repair," or "scrap." Make sure "scrapping" renders the item unusable, so it cannot be used mistakenly and cause an accident.
2. Store reusable equipment carefully, so it can be unloaded safely.
3. Make sure the tags and locks are all removed from site.
4. Make sure the client is aware of demobilization and has the company forwarding address and a contact telephone number.
5. Make sure all hazardous materials and chemicals are disposed of carefully, correctly, and confirmed to the client as having been removed.
6. Leave the site clean, safe, and free from debris.
7. Ensure that all items left on site are clearly marked.
8. Now that the project is complete, it is very important to ensure that all the safety records are returned to the corporate office:
   a. so that the corporate safety office has a chance to analyze and record the data for historical, federal, and state information purposes
   b. so that the experience gained on this project can be reused and recalled at any time, to be incorporated in the next project

4

The Safety Program "Up and Running"

All the individual pieces of a good effective safety program have been accumulated and explained. Now we have to put them all together to see how this, at first sight, immense amount of extra work can all be accomplished without six extra people on site.

We have basically five different groups of activities which have to be melded together: inspections, reporting functions (paperwork), meetings, research planning and development, and program self-evaluation. All these activities somehow have to be put into an already cramped schedule and be handled by somebody on the site staff. How can it be done? It's impossible, some would say. They would be wrong.

The only way I know of showing that it is not impossible and that it is relatively simple is by inventing a job site and writing about a hypothetical project from bid to completion.

The hypothetical project is the building of an outdoor overhead crane runway extension. It does not matter where. All the information is based on facts accumulated in a typical state and by the federal government, and all the facts used are easily accessible to the general public.

The site engineer has been appointed to become the safety coordinator by the site manager. Being a very well-prepared chap, this site engineer brought with him, when he came to the site, a lot of safety materials to tide him and the site over until an order could be placed to purchase all the necessary additional safety equipment and supplies required based on the size of the job, manpower, and conditions.

His first job as safety coordinator is to review the bid package with which he is already familiar since he has been on site a couple of days and has already started to get accustomed to the scope of work, etc. He makes notes on the safety information included in the bid documents, including the client's demands. This takes approximately 2 hours, with the following results: Safety has a budget in the estimate of 1 percent of the total project value of $2,820,000 which is not very much— $28,200—but it will have to do as the bid on this job was extremely tight.

The heaviest columns weigh 12 tons and the largest lift is the overhead crane and bridge girders at 24 tons each. No hazardous chemicals of any extraordinary nature are envisioned to be required on site. It is straightforward construction: remove existing shed, excavate, drive piles, pour foundations, erect steel columns, building structure, pour concrete floor area, erect overhead traveling crane and commission crane, clean up, and finally leave site.

Project duration is 6 months. Manpower, largest total, is approximately 40 workers. Crafts to be used: laborers, carpenters, operating engineers, ironworkers, millwrights, electricians, and pipe fitters. So we know we need craft rules and regulations for all of them. Numbers of each craft will be forthcoming as the bar chart and critical path schedule are established.

At the prejob meeting with the union representatives, all special conditions and general conditions have been established. The meeting minutes are in the file, and a quick call to labor relations at the head office produces not only the latest information but a copy of all the client's requirements too. All these relevant rules and regulations and general and special conditions are now put into some presentable format and sent to the printers to make up durable (plastic-coated) "easy-read" cards that anybody can carry in a top pocket and refer to when required.

Since the work force is to be relatively small, job orientation can be very easily and economically accomplished verbally, without a film or video. The safety coordinator brought with him all the forms and filing system requirements, so the job secretary has already got files and folders typed up and organized. He also brought a good selection of craft rules and regulations cards with him. He does not have quite enough so he sends back to the head office via an internal memo to get some more sent out. The work force first hiring will begin next week, so all the paperwork required for the first job orientation has to be on site ahead of time.

Just as he finishes up this basic review, the first delivery truck with job supplies rolls up. This shipment is an internal transfer from another job site in the same area. This load contains fire extinguishers of all types, hard hats, safety glasses, hand tools, safety belts, emergency signs, and many other useful items. The safety coordinator stops the mad scramble to raid the goodies and insists that the site clerk, who arrived on site at the same time as the site manager and the safety coordinator/site engineer, inspect everything first to ensure that all is in good safe working condition, *before* being distributed.

In the following few days, by devoting another couple of hours to the safety program, the other parameters can be established. The job orientation presentation and the various handouts to each site employee, e.g., hard hats, ear plugs of some sort (for noise protection), safety glasses (the client has demanded that everybody on the job site wear safety glasses with side shields while in the field), plastic-coated craft regulation cards, are all available. Locations of fire extinguishers, emergency plan information, telephone locations, and all other relevant information is itemized on the plastic-coated, printed job-site regulation cards. Emergency information is posted wherever necessary. Relevant toolbox safety topics have been accumulated from various sources (the insurance agent, OSHA, and the company head office). The job is now ready for its first influx of craft personnel. This happens the following morning.

Because of the planning and coordination, everything goes very smoothly. The clerk signs them up, and then they proceed to the conference room (a partitioned area of a job-site trailer) where the safety coordinator reviews all the necessary information. They sign the orientation meeting sheets, and they are ready with all their handouts to start work, properly attired. This takes 30 minutes. As everybody gets more organized, each group of orientations take less and less time, until the whole orientation program takes 20 minutes or less.

The most vital of all the important topics covered in this orientation is the manner in which the safety program works: the inspections, the safety committee, the craft representatives, the suggestion box, the evacuation plan, the corporate safety commitment flyer, and so on.

The next order of business is to get the safety committee organized and the committee members prepared. (Initially use the craft foreman or the union steward, as previously suggested in an earlier chapter.)

Each craft representative spends 30 minutes with the safety coordinator reviewing the requirements of the safety committee. He receives pertinent toolbox safety topic review sheets (Figures 4.1 through 4.8), a job-site inspection notebook, and a list of the most common accidents which occur in his particular trade. These have been prepared by the site engineer, using the information available from the corporate office, the Bureau of Statistics in Washington, and each union headquarters of the crafts involved on the project. (See Chapter 5 for more details.)

A good clout on the head by a piece of falling material is one way to convince a holdout that he needs a hard hat. But that's a bit drastic, and in some cases, permanently damaging. We'd rather try and answer the arguments these workers usually give when asked to wear a hard hat on the job.

"It's too heavy." Hard hats are a few ounces heavier than a cloth cap, but the extra protection is worth the extra weight. And, a hard hat is less than one-third as heavy as an army helmet.

"It's too hot." Actual measurements have shown that the temperature under a hard hat is 5 to 10 degrees cooler than outside.

"It gives me a headache." A thump on the head from something which has fallen two floors will give you a worse one. However, there is no medical reason why a properly adjusted hard hat would cause a headache.

"It won't stay on." You're right it won't. Not in a high wind, anyway. But a chin strap will solve that problem. Otherwise, you'll find that the hat stays put no matter how much stooping or bending you have to do—if it's fitted correctly.

"It's noisy." That's your imagination. In fact, tests show that properly worn hard hats will shield your ears from noise.

If you stop to think about it, the hard hat is a very useful piece of safety equipment.

QUESTIONS FOR DISCUSSION

1. What are the adjustments that can be made on our hard hats for proper fit and comfort?
2. Why is it important that hard hats be worn all the time by everyone on the job site?

Figure 4.1 Toolbox topics: Hard hat or hard head. *(The examples shown in Figures 4.1 through 4.8 have been provided by the kind permission of Liberty Mutual Insurance Company, Boston, Massachusetts. This company provides many other construction safety topic discussion sheets such as these.)*

One of the most used, often abused, and least noticed pieces of equipment on the job presents a major hazard. This is the ladder.

Out of 150 construction accidents involving ladders, it was found that the following were principal contributing factors:

1. Climbing or descending improperly
2. Failure to secure the ladder at top and/or bottom
3. Structural failure of the ladder
4. Carrying objects while climbing or descending

Generally speaking, commercial ladders or job-built ladders are constructed properly and are of sound material. However, after they have been in use for some time, they are often damaged through abuse, rough handling while moving, being struck by heavy objects, etc. Failure on the part of anyone using a ladder to report a defect may result in a serious fall.

You will hear many arguments about the best way to climb a ladder. Many people say, "Use the hands on the rungs"—still others say, "Grip the side rails." Most people agree that either method is okay if you use both hands!

Too often ladders are not secured either at the top or the bottom. It takes only a few minutes to tie-in a ladder—it takes a lot longer to heal a broken leg, still longer for a broken neck, and, no matter how long you try, you can't restore life.

Ladders should be set at the proper angle: fixed ladders between vertical and 75 degrees; others should be set out at the foot a distance of about one-quarter the length of the ladder.

Ladders should be long enough to extend at least 3 feet above the landing.

When it is necessary to get tools and/or materials from the ground up to the work level or down again, don't carry them on the ladder—use a hand line to haul them up or let them down.

QUESTION FOR DISCUSSION

1. Are the ladders on this job in good condition, and are they properly used?

Figure 4.2 Toolbox topics: Ladders.

I'm sure you all know that falls cause more injuries in the construction industry than any other type of accident. In fact, about 40 percent of the serious injuries in the building trades are due to falls from one level to another.

Good protection of floor openings is one way of preventing these falls. This protection is a responsibility of management, but it is also a responsibility of the trades. If you have to remove guardrails or covers to work or hoist in a shaft, put the protection back when you are done.

In one recent accident, two laborers were cleaning up a floor area. They piled scrap lumber on a sheet of plywood and then picked up the sheet to carry the material away. Unfortunately, the plywood had been covering a floor opening and the rear man walked into a 25-foot fall. When you cover a floor opening, secure the cover so that it won't be moved by accident.

If you remove a section of a steel grating floor, rope off the area. These openings are particularly hard to see when the floor below is also steel grating.

QUESTIONS FOR DISCUSSION

1. Do you know of any locations on this job where floor protection is either lacking or defective?

2. What procedure do the mechanical trades follow on this job for replacing or arranging for replacement of floor opening protection after they have removed it?

3. How do we draw attention to these hazardous areas and make them highly visible?

4. Why is it necessary to clean regularly around these hazards and keep them clean of stored material?

Figure 4.3 Toolbox topics: Floor openings.

So the project moves into full swing. Carpenters, electricians, operating engineers, and laborers are the four trades which are on site at this time.

The areas of most concern should be proper trenching and shoring, ground fault circuit interruptors, good safe routing of temporary power, lumber and plywood with old nails bent or removed, careful

A carpenter asked his insurance company to pay for the damage done to his glass eye which was broken when a nail he had been driving flew up and struck it. When he was asked how he had lost the eye in the first place, he replied, "Oh, the same way, a nail flying." A world of darkness awaits this man if sometime the nail hit his only good eye because he has not yet decided to take advantage of the protection of safety glasses. It may be difficult getting used to eye protection, but have you tried getting used to a glass eye?

There are two kinds of foreign particles which get in your eyes on construction jobs. One kind is the material carried by the wind— sawdust, flecks of ironrust, etc. These aren't too troublesome. Then there are the high-speed chips which result when one hard material comes in contact with another hard material. Here is where you really need eye protection:

1. Driving hardened nails in concrete

2. Running a jackhammer to break rock or concrete

3. Drilling, reaming, or chipping metal

4. Hammering on a chisel or a steel pin

QUESTIONS FOR DISCUSSION

1. What are the jobs we do where eye protection is needed?

2. Do we have proper eye protection available and is it used?

Figure 4.4 Toolbox topics: Eye protection from falling particles.

There are many, many ways we can hurt our backs. Let's discuss two case histories. Perhaps if we know more about how they happened, the same thing won't happen to us.

A mechanic was searching the scrap pile for a piece of steel angle he could use for a repair job. Seeing none of the right size around the edge of the pile, he climbed up on the pile. He spotted what he thought he would need sticking out from under a 6-inch I beam which was about 4 feet long. He jerked one end of the beam and felt a severe pain in his lower back.

A laborer was breaking ground in preparation for the installation of a service line. He was operating an air hammer, and the point of the tool became wedged in a short length of 2×4 which was frozen in the ground. When he attempted to jerk the tool loose, he suffered a muscle strain in the upper back.

Here are two cases where active, physically able men hurt their backs.

QUESTIONS FOR DISCUSSION

1. Both of these men were doing the same thing when they felt the pain. What was it?

2. How would you have done each of these jobs safely?

Figure 4.5 Toolbox topics: Hurt backs.

Contacts between crane booms and power lines cause more fatalities each year than any other type of electrical accident in the construction industry. If you are not personally familiar with one of these accidents, you certainly have seen them reported often in the newspapers.

Let's review several typical accidents involving these contacts.

1. An ironworker had hooked onto a bundle of rebars stored under a power line and was guiding the load when the boom hit the line.

2. A pile driving foreman was walking backward pulling the hook when the load line contacted an overhead power line.

3. An oiler was leaning against the side of the crane when the boom hit a power line causing the current to go to ground through his body.

Each of these accidents resulted in injury or death to someone other than the operator. In fact, the operator is usually safe when a contact is made and is able to protect himself either by swinging the boom free of the line or jumping clear of the crane. It is the man on the ground who gets the electric shock. Both the man on the ground and the operator should realize that it is difficult for the operator to be sure of the exact location of the boom tip. The operator just doesn't have good distance judgment looking up along the boom, and he is usually pretty busy watching the load anyway.

The best way to avoid contacts is to keep the boom at least 10 feet away from any overhead line. This may mean storing material in some location that is less convenient than the empty ground under the wires. It may mean that someone has to be assigned to watch the boom tip when work approaches a power line. It may be necessary for the power company to deenergize a line or protect it with rubber sleeves.

Finally, if a man does come in contact with an electrical source, don't try to pull him free with your hands. If you have to free him, use a dry manila rope or a dry plank.

QUESTIONS FOR DISCUSSION

1. Do we have any material stored or work to be done close to a power line on this job?

2. Is there any one here qualified to give artificial respiration to a man rendered unconscious by electric shock?

Figure 4.6 Toolbox topics: Watch the wires.

We all want to keep from getting hurt on the job. Each of us has a responsibility to look after the "other guy" who may follow along after us or use the same tools, equipment, or material.

We can't pass the buck about safety because it isn't a one-man job. Never say, "I'll take care of myself and let the other guy take care of himself."

If one of the other men is working in a dangerous position, warn him about it. He may not be experienced enough to recognize the hazard, or he might have problems that are distracting him.

Worring about being considered a "wise guy" should not keep you from offering advice on safety. Advise in a helpful, sincere way and your interest will usually be appreciated.

Here are some ways we can help each other work safely:

1. Set an example in the safe method of using tools and equipment. Help the inexperienced man learn the right way.

2. Keep machine guards in place, and don't leave a trap for the "other guy."

3. Report machine defects or accident hazards to your supervisor promptly.

4. Encourage everyone to report for first aid at once for every injury no matter how slight.

5. Encourage the wearing of proper clothing and personal protective equipment.

6. Ask questions if you don't fully understand your job.

QUESTIONS FOR DISCUSSION

1. Do we always report unsafe conditions that could catch someone else unaware?

2. Can we take safety suggestions in the cooperative spirit in which they are made?

Figure 4.7 Toolbox topics: How can I help reduce injury to others?

Over one-third of the serious injuries to men in the building trades are caused by falls from one level to another. Think about the falls you have seen or heard about.

These falls usually occurred because the injured man did not have a safe place to stand while he did his work. You probably can recall accidents when the man set up his own makeshift scaffold or used some convenient pile of material because he didn't want to take the time to do the job right.

If you don't care what you work from, almost anything will do. A pile of concrete blocks or even cardboard boxes will hold you—if they don't tip over. A single sloping plank supported on one end by a pipe and the other by a stepladder will put you where you can do the job—provided that the plank doesn't slide or you don't step back. The curving metal top of a blower or tank will do—if your foot doesn't slip.

The time and the materials are available to build a safe scaffold for each job, and a good craftsman knows how and when to use them.

QUESTIONS FOR DISCUSSION

1. Why do we see so many makeshift scaffolds on construction jobs?

2. Is the time used in setting up a safe scaffold saved by providing a place where a man can work without worrying about every move he makes?

Figure 4.8 Toolbox topics: Makeshift scaffolds.

barricading of the swing both of the crane and crane cab (both should be reviewed), and last but not least, storage of equipment away from interferences with work areas or personnel.

The daily inspections take place as required by the safety program, and the project and the safety program are working smoothly. Weekly safety meetings consist of reviewing site conditions, which takes a few minutes; then the most important two topics on the agenda are dealt with. The first topic is coordination of the next few weeks' work plan, ensuring that crews and machines are geared up (safety-wise) for the following week's activities. Craft training requirements have to be discussed: Are there enough competent first aid people and any CPR-trained people on site? Are there any new conditions on site requiring more craft awareness? Are the crafts integrating properly and compatibly? Is there any unnecessary friction?

Somebody tables a question about additional toolbox meeting hazard awareness literature, which the safety coordinator agrees to find. The general superintendent has seen some good safety training films on a recent visit to the head office and suggests that they are shown at site. Motion carried.

The second important topic to be covered is the predictable hazards to be encountered during the following few weeks' activities. Ironworkers are going to be on site shortly. Steel is going to be erected. What is the criteria for site safety for this, the most hazardous activity during this project? Safety belts will be worn. Overhead work is going to interfere with the concrete activities and underground conduit installations, so should safety nets be used? Discussions follow.

The safety coordinator now presents activity hazard analysis, explains how it works, and delegates two AHA reviews to be done: one on the steel erection and one on the crane erection. Further discussions take place on steel: Is steel going to be stockpiled for easy access and the minimum of interference to the other trades? The project schedule is reevaluated so that foundation and concrete work do not impede progress of steel erection, remembering that each time a steel lift and connection above ground takes place, there must be no personnel underneath the crane boom or close to its radius of movement.

This whole meeting takes approximately 1½ hours. The total time spent to date by the safety coordinator per week, to this point, is 6 hours per week, or just over 1 hour per working day. The coordinator's job is made up of research and planning, job-site orientations, job-site inspections, documentation, and safety meetings.

In addition, the engineering clerk has spent 1 hour each week making sure that the fire extinguishers are in the proper place, that they are fully charged, and that all the necessary emergency and information paperwork is correctly displayed in the correct place, e.g., craft trailers, by each of the telephones etc. The clerk also makes sure that danger signs are posted by hazards.

The general procedures of the safety program are now fully established, and everything starts falling into place (to use a most inappropriate expression). The site operations are smooth. The only "hiccups" to date have been two cut fingers which were minor and required only first aid treatment by the craft superintendent (both marked down to personal carelessness) and dirt in a carpenter's eyes during a particularly gusty wind (no hospital treatment was necessary). The accident was caused by working too close to an earth-moving activity, but basically by lack of communication between the carpenters, the operating engineers, and the laborers. All three incidents were discussed in detail at a safety committee meeting, and simple steps were taken to avoid a recurrence. Resolutions were, of course, recorded in the safety meeting minutes. All pretty straightforward and simple so far.

We are 8 weeks into the project, and all is as it should be. The safety budget is still in the black, and the estimated final cost is not forecast to exceed the budget figure. Remember, none of the work force's time is being charged to the safety budget as these activities are part of normal project work and should be charged to each direct discipline of work. It is all part of executing each activity correctly and making sure the workplace is consistently safe. Similarly, the safety coordinator's time should not be charged to the safety budget—since that time is an integral part of the normal indirect costs of company site management.

The only items charged to safety are the safety equipment, hard hats, safety glasses, safety nets, literature, and safety consumables. That is why the program is within budget. On a site where there is a full-time safety coordinator, budgeting is different, but of course there will be a bigger budget in the bid documents to reflect that.

Now the fun begins.

Into the office comes the ironworker supervisor holding two very sick portable grinders. A quick inspection shows them to be burned out, and their plastic bodies have melted. They have been working for only 5 hours, the supervisor informs the safety coordinator, and 6 grinding wheels have disintegrated. Further research shows two problems. The faulty grinders both have similar malfunctions. The requisition from the field to the office shows that heavy-duty grinders and grinding wheels were specified. The purchasing agent had been offered a cheap job lot on sale, and thinking he was getting value for money, he placed the order for the cheaper models.

Fortunately, there were no injuries. The wheels did not do any damage when disintegrating, and there were no electric short circuits, so no damage was done. But, oh, how close. The gods were smiling.

The safety coordinator now approaches the purchasing agent to ensure that the incident is fully explained and the same thing does not happen again. The poor purchasing agent is not very happy that his efforts at economy have been wasted, but he has learned a vital lesson. He, of course, blamed the heavy-handed ironworkers, but common sense prevailed, and he did not say that to them! Tragedy has been averted, and new heavy-duty grinders are purchased. Calm, orderly business resumes.

The next highly unusual event is at the end of the workshift the next day. The craft trailer for the laborers erupts into pandemonium. The office is still working when in rushes a laborer to say that a man has collapsed. The safety coordinator delegates his clerk to phone an ambulance (a 911 call) and goes to see the problem for himself. He finds everything pretty much under control. A carpenter from the adjoining half of the craft trailer has stepped into the breech and is performing CPR. The patient is now breathing, the carpenter's mate (also trained in CPR) is assisting, and between them, they keep him breathing until the two paramedics appear and take over, rushing the patient to the hospital.

The safety coordinator now takes stock of the situation, informs the site manager of the incident, and then starts an accident review. Where was the man working? How old is he? Did he complain to any of his mates? Was he sick? The review is completed as best it can be at that time; the rest will wait until the next morning. The safety coordinator checks with the site manager to ensure that all the proper people have been informed, and he finds out from the hospital that the man is surviving, is in intensive care, but is still critical. The family has been informed and is at his bedside. Nothing more can be done except to leave a contact number in case of further developments.

The following morning, the site inquest begins. It is established that the man did complain in the early afternoon of being "out of sorts," but he also said he had not been sleeping well, so he figured an early night would solve his problems.

It is also established that the two carpenters saved the man's life by quick efficient action. This is brought to the attention of the site manager. He decides to take this matter further to see if some reward may be made to the two men. Predictably, the two men refuse the offer—it is part of the job, they say. The man survived, and that is thanks enough. The true craftspersons, the salt of the earth, have come through with flying colors again.

At the following weeks' safety meeting, these two diametrically different incidents are reviewed in detail by the committee. Two motions are adopted: one, to nominate the two carpenters for a community first aid award, and two, to inform the purchasing department at the head office and on site of the near-miss with the cheap grinders.

A bad week, yes, but still the safety coordinator has spent only 9 hours on the safety program, less than 2 hours per day that week.

The excitement of the previous week dims, and the project returns to normality. This week everything goes so calmly that the safety coordinator spends only 3 hours on safety activities.

The project is in full swing, and all crafts are on site. Craft forces are peaking, and the safety program is now devoting more and more of its time to activity hazard analysis, ensuring that communication occurs at all levels effectively and that the job site is kept clean and tidy, which means less opportunity for an accident to occur. However, by delegating a lot of the committee activities to the committee members, the safety coordinator is now spending 4 hours per week on safety, of which 1½ hours are taken up with safety inspections. The rest is spent in his site engineer's duties.

The project is 12 weeks old, completion is close to 40 percent, and the first safety program audit is due. This audit is to be undertaken by some head office personnel who nobody knows. So the first job is to ensure that everything is in relatively smooth order before they arrive.

The safety coordinator and the site manager meet and formulate a plan of action. They delegate the site clerk and the subcontractors' administrator to review the program before the audit. In the meantime news of the audit is related back to the safety committee. Some of the review activities are delegated to them while the paperwork is briefly reviewed by the safety coordinator before being presented to the self-evaluation team. The internal review takes place, and the only negative finding is that the files containing the craft toolbox meeting minutes are not complete. There are some holes in the sequence of weekly minutes on several crafts.

Some hurried consultations with the responsible craft delegates result in the finding that the safety coordinator has misfiled the offending reports, and all is put back into good order.

The next week the audit team arrives and, using the audit forms shown in the appendix, reviews the safety program—complete with entrance and exit interviews—within a 9-hour day. With the exception of one or two minor glitches, all is found to be in good order, and everything proceeds very successfully.

With all this activity in the preceding 2 weeks, the time allocated by the safety coordinator to the safety program is 10 hours per week. The overall average of the whole job so far is still 6 hours per week.

Yes, the craft representatives spend several hours a week on safety, but even without a program, this time would have normally been spent on the same subject, but, of course, it would not have been well planned or coordinated. It would have been haphazard and far less productive. So, basically, as you can see, the program takes no extra personnel—it requires only that the same personnel be used more effectively.

The project is now in week 15 with everything still moving smoothly, but distinct rumblings are afoot. The project gathers momentum, and the ironworkers are really fighting to stay on schedule. The columns are up, and the connecting beams and bracing are up. Now the roof steel starts, and the fight begins.

It will save a great deal of time, argues the ironworker superintendent, if we do not have to put up a safety net. We, ironworkers, do not need one. It destroys our "macho" image for one thing. We do not fall anyway.

The following week's safety meeting has this topic at the top of its agenda. The committee has already heard that the ironworkers do not want to use a safety net. Everybody on the project is very aware of the ironworkers' outstanding safety record so far, and understandably, they do not want to lose it. They had been more or less in tentative agreement earlier in the project that there would be safety nets. The mood of most of the safety committee members is resolute and determined. Now watch out for fireworks.

The safety meeting starts with a review of previous meetings; what was said, by whom, referring, of course, to the historical meeting minutes which have been resurrected for this meeting. The ironworker superintendent makes a good case as far as its schedule is concerned and the initial outlay of effort for no actual production. But the labor supervisor, with no real axe to grind, points out the false value this reveals. Regardless of how much effort is expended in putting up the net, if it saves one accident, it is worth it. There is no answer to that reasoning. After a great deal of heated argument, common sense prevails. The ironworker superintendent compromises, and safety nets will be installed. This exercise has taken 1 whole hour, but it is important to note that nobody has been embarrassed during the discussions, and everybody eventually leaves the meeting, after the other business is expedited, with good positive feelings. The right path has been followed. Time alone will show if the decision was worthwhile. One of the principal factors in the argument which swayed the decision was the fact that people can work (with extreme care, of course) underneath the safety net, thus allowing other disciplines of the project to forge ahead and, in fact, overhauling the schedule in some areas and somewhat compensating for the extra time with no production reward while erecting the safety net in the first place.

The extra research by the site safety coordinator for this meeting, and the length of the safety meeting, mean that this week he has spent

12 hours on safety, but the average still works out to 6 hours per week, just over 1 hour per day on the project overall.

So, the project progresses, and I am sure you can imagine what happens next. Yes, the safety net claims its first catch—not a human, but a piece of equipment. A torque wrench, weighing several pounds and costing several hundred dollars, falls. Because of the net, the wrench suffers no damage. The carpenter walking underneath this area when the wrench fell never felt better, knowing that without the safety net, his wife would now be entitled to claim his life insurance principal sum. So the net has already saved an incalculable sum of money and possibly one person's life. It also happened to be the only torque wrench on the job. Delivery of these wrenches takes 2 weeks, and, had it been damaged, the lost production would have been very detrimental to the schedule. The project completion date would have all of a sudden been in jeopardy.

I think the point is made; safety nets, no matter what the original outlay is, are very worthwhile. Just ask a few old craftspersons what they fear most when working. Yes, almost invariably the answer is something dropping on them from above. If you doubt, ask away the next time you are on a construction site.

The project is now at full steam ahead, and the ironworkers are finally gaining on the schedule—and still no accidents. An occasional bruised set of knuckles, more dirt in more eyes. One chap had dirt from one of the overhead beams fall in his eyes, and he had to go the hospital to have his eyes washed out, but he was released an hour later and returned to work the next day. Therefore, there was no lost time reported.

As you will see in the next chapter, and we digress again for a moment, the full benefit of the client's unpopular regulation to wear safety glasses on site has avoided many more serious injuries. One of the most costly and regular occurrences on a job site has been significantly reduced. The safety coordinator notices this every time he fills out the monthly OSHA 200 report. He decides to write a memo to the head office hailing the necessity of having to wear safety glasses with side shields and that eye injuries have been reduced by 80 percent based on records he has been shown by the site insurance representative. He feels quite rightly maybe everybody on construction sites should wear them. Oh, what a giant step of progress that would be for construction safety. I can honestly attest to the fact that, never having had to wear glasses, it is a bit uncomfortable at first, but after several months of perseverance, they become unnoticeable and second nature.

Anyway, enough philosophy, let's continue with the project in full flight.

The safety meetings, inspections, AHAs, and reporting continue without incident. The safety coordinator is back down to 4 hours per week spent on safety, and all is quiet on the project; first-class productivity, a clean tidy site, and no lost-time accidents. There have been no more problems with tools, and the safety net has paid for itself several times over.

The safety coordinator, 20 weeks into the schedule, makes one last purchase order to last him the rest of the job for safety-related supplies and equipment. The order is minimal, but it takes him $250 over the budget estimate. Now the site manager gets a little perturbed because the figures by which his mentors review the job is only by cost and budget estimates. The cry goes up that the safety program is being extravagant. Hold it right there, says an irate safety coordinator. Let's evaluate that statement and the facts. He's got both barrels ready.

1. There have been no lost-time accidents, so the insurance rates will go down on the next job.
2. There will be a refund on rates for this job due to no lost-time accidents.
3. The budget estimate for lost time was 5 percent of the direct project costs.
4. Production costs are down due to greater efficiency due to a clean, safe site.

All right, all right, you've made your point—don't go on, says the site manager. Consequently, the purchase order is approved without another word.

The next area of business is getting back to the first order of priorities. The project is getting ready for the most important part: commissioning the electric circuits and the overhead crane. Notice that there has been no mention of the subcontractors to date. There are only three subcontractors: the electrical company, the painters, and the concrete floor finishers who are laying a hard-top screed on top of the floor. All the rest of the work has been done by direct hire. Believe it or not, a subcontract administrator has been on site since the electricians have been on site, a week after the project started, and they have all been a pleasure to work with, even the painters who are notoriously difficult to coerce into towing the policy line. They have all been diligent in their attitude to safety and site cleanliness, good housekeeping, and attending the safety meetings regularly. Impossible, you may think. Well, it does happen when strict observance of the contract documents prevails and negotiations on site are firm but fair. The subcontract administrator has done his job outstandingly and has, in addition, helped out the safety coordinator on occasion. He has been a boon to the whole management of the project. That is why subcontractors have not been mentioned before, because it is all going so smoothly. I said this was a hypothetical project! No, it is not the perfect project, but the work ethic has been so good by everybody concerned that, yes, it is close to perfect. Some projects do run like this hypothetical one when people are dedicated to making the effort to plan and execute the safety program with deliberation and concern for the rest of their colleagues. Then it quite naturally becomes a pleasure to go to work. It is all part of being a professional in your own field. There are many, many people like this in all trades and in site management, and if given a chance to plan and manage project execution, rather than manage from crisis to crisis, many other projects could be this way too.

So back to the commissioning phase. The safety coordinator and site manager, in one of his infrequent visits to the safety committee meetings, presents the method of operation and procedures for starting-up a piece of equipment or energizing an electric circuit. The rules and regulations are established, and everybody understands clearly his or her own responsibilities. There will be meetings just prior to the finish of work each day to enumerate the following days' activities for the duration of this phase of work.

Commissioning has some "hiccups," and just like the early days of the project, it takes a week before everybody involved becomes a finely tuned team. But without any major incidents, progress becomes the ultimate winner, and all the equipment eventually works just as it was supposed to. Two weeks early, the project is turned over, in full working order, to the happy client, who, incidentally, has not said much for the whole job. He has been amazed at how smoothly everything has worked out and has already recommended that the same company be sure to get the next job at a sister plant getting the same plant extension in 6 months' time.

The construction team has completed the project under budget and has made a handsome profit. The whole team should be congratulated. The results have also had a good effect on the head office estimators, for they now know they are on the right track when estimating similar jobs. An additional bonus is that the insurance rates for future construction by the company have been significantly reduced, providing other jobs are run as this one was. This is a direct result of the success of the project.

The final task before leaving the site with chests puffed out is packing up all tools and equipment, scrapping the damaged tools and equipment, making repairs, and carefully storing for transit everything that is going to be moved onto the next job.

The client makes a final inspection, a forwarding address is left, an emergency telephone number is passed out, and a final lunch is had with the client before the "good-byes and good luck."

The safety coordinator has fed all the final project safety figures back to the head office, together with all the pros and cons of the program, and with one final walk around the site, drives off to the next challenge.

His time on safety worked out to under 6 hours per week, on average, for the whole project. Small outlay for such outstanding success. Try it. I guarantee it will produce results if you do it properly with diligent people and the support of both the trade unions and your company management. A joint labor-management effort is what it takes to make the program work.

The whole philosophy of this project has been, in retrospect, getting everybody involved from the site manager to the teaperson. There is an inherent need in individuals to feel that they are participating especially when whatever they are doing is successful.

It is very worthwhile to "walk the extra mile" to ensure that everybody does participate in the whole process of the project. Notice that on this hypothetical project there were no grievances mentioned. There was a good reason for that. On a successful project, there are no grievances worth mentioning because everybody feels good. They feel good because they have had a satisfying work experience. The quality of working life is good.

Job satisfaction breeds success in all facets of work. So it is with a good construction project, and so it is with a good safety program.

5

A Few Interesting Facts

The purpose of this chapter is to highlight some little-known information which has become apparent during my research in designing and executing safety programs and in writing this book. The first series of facts are the basic causes of the most common accidents occurring in the general trades in construction. Again, this information comes from current national and state statistics.

Common Causes of Accidents

Remember: If an accident is predictable, it can be avoided.

Laborers

1. Overexertion
2. Being struck by objects
3. Being struck against objects

Poor trenching is still a major cause of the loss of lives, in most cases because of inadequate shoring.

Carpenters

1. Overexertion
2. Being struck against objects
3. Being struck by objects

Exposed nails in discarded lumber and hand tool failure are the cause of very high incidence rates. Scaffold failures are still the cause of considerable damage. Most of these are predictable accidents which can be avoided.

Operating engineers

1. Being struck by an object
2. Being struck against an object
3. Losing foothold

Machinery failures (mostly predictable failures which could have been avoided with diligent, regular inspection) cause the majority of serious accidents.

Ironworkers

1. Falls from elevations
2. Overexertion
3. Being struck by an object

Serious finger injuries are very common—getting fingers caught between two objects.

Painters

1. Falls from elevations
2. Being struck by objects
3. Skin poisoning

Paint sprayers are a large cause of many unnecessary, unintentional accidents.

Millwrights

1. Overexertion
2. Being struck by objects
3. Caught between objects
4. Eye injuries

Serious finger injuries are very common with this trade also.

Electricians

1. Overexertion
2. Being struck by objects
3. Electrical contact

High-voltage electrical contact is by far the most dangerous of the electricians' hazards because the least resultant injuries are serious burns; the worse, of course, is fatality.

THE LONG STEP!

HOLD IT! JOE ~~~ WHY NOT TAKE A STEP IN THE RIGHT DIRECTION, BEING SAFE WON'T HURT A BIT.

BLANK

ALWAYS BE CAREFUL, USE GOOD JUDGEMENT AND PRUDENCE IN DOING YOUR WORK, SO AS TO PROTECT YOURSELF AND OTHERS FROM INJURY, WHETHER OR NOT THE ACTS NECESSARY ARE INCLUDED IN THE SAFETY CODE.

Masons

1. Being struck by objects
2. Overexertion
3. Falls from elevations

Using poorly constructed scaffold is a major cause for predictable accidents to occur.

Boilermakers

1. Overexertion
2. Being struck by objects
3. Falls from elevations

Temporary scaffolding and work ledges are certainly a high risk for this trade.

The above facts are all common sense and generalities, but they are worthwhile to emphasize in your own project when looking for subjects for safety training.

Sources of Accidents

The next group of facts highlights the main sources of accidents in the field, which obviously need the greatest attention during the safety program.

Scaffolding and staging
Powered conveyors
Electric and mechanical rotating equipment
Powered hand tools, especially saws, grinders, hammers, and drills
Hoists and cranes, chains, ropes, and cables
Elevators and elevator shafts
Jackhammers
Wooden ladders
Saw machines
Trenches
Discarded lumber

Opensided floors

Housekeeping

Poorly supported temporary power

There is no particular order of priority to this last list. All items listed produce their own share of catastrophies and the ensuing lost work-days and, in some cases, unfortunately, fatalities.

Safety Aids

Some of the most important aids to safety which have a great effect on reducing the lost-workday accidents are:

Hard hats

Safety glasses with side shields

Gloves

Ear defenders

Safety belts

Catinary line life-support systems when working at high elevations

Safety nets for high-elevation work

Tied-off metal ladders

Safety shoes

Properly designed gas-cylinder carts

Having a good percentage of properly trained first aid and cardio-pulmonary resuscitation (CPR) craftspersons on site (If there aren't, then make the effort to hold training courses to encourage it.)

Then, of course, there are the intangibles which are designed to shock the complacent people on the job site out of their lethargy. Safety films of all types are available free from all sorts of sources. They are great media for making a person think about his or her work in a different context, and you would be surprised how effective they are in helping to persuade people to exercise a little more effort toward personal safety.

ALWAYS LOOK AHEAD

HEY BOSS~ WHERE'D YOU SAY THAT OPEN SHAFT YOU WANTED COVERED WAS ???

YOU CAN'T MISS IT. GOOFINOF !!!

LARRY WEINMANN

IF YOU HAVE OCCASION TO REMOVE THE COVER FROM ANY OPENING IN THE FLOOR, ALWAYS GUARD THAT OPENING SO NO ONE WILL BE IN DANGER, ALWAYS REPLACE THE COVER WHEN YOU ARE FINISHED.

Safety Hazards

In addition, here is a list of site hazards and basic protections to which particular attention should be paid:

Barricades

Use these around all the following:

    Opensided floors
    Trenches
    Floor openings
    Rotating equipment
    Crane cab swing radius
    Elevator shafts

Sight screens and good ventilation

Apply these in the following areas:

    Welding in confined spaces

Paint spraying activities

Machining operations

Here are some other suggestions or safety criteria which could prevent an accident:

Cranes

Only take instructions and signals from one person. Make sure he or she uses the proper signals.

Keep the crane on a firm and level surface. When leaving the crane, set the brakes.

Use care to avoid a whipping action with the load on a long boom.

Concrete work

Avoid working directly above protruding rebar unless it is covered and hence will eliminate impalement.

Safety glasses and/or face shields should be worn when chipping concrete or using power tools on concrete.

Do not stand underneath a concrete bucket or underneath the boom lifting or moving the concrete bucket.

Use protective clothing when working with wet concrete as it has chemical properties which are hazardous to the skin.

Keep walkways clear of spilled concrete.

Do not climb on concrete forms.

Protruding nails and wire ties and other accessories should be removed or protected as soon as the concrete forms are stripped.

Wires under tension should never be cut until the tension is released. This will avoid dangerous backlash.

Cutting and welding

Avoid placing gas cylinders in confined spaces.

Separate the oxygen bottle from the acetylene bottle.

Ensure the gas cylinders are stable and restrained at all times in an upright position.

Oxygen and grease do not mix. It can lead to explosions and serious injury.

Gas cylinders when in use should be checked regularly for leaks.

Make sure, before cutting, that hot metal cannot be the cause of starting a fire in the immediate area.

Pipe fitters

Pipes should always be stored horizontally and blocked to prevent movement.

Do not walk on pipes.

Pipe wrenches and other pipe fitters' tools can slip on the curved surfaces. Make sure the tool teeth are sharp and clean to avoid their slippage.

When working on operating systems, *ensure* the system is locked out *before* working on the system.

Assume all piping has a fluid or gas in it until it is proven otherwise.

In an operating plant, never touch a valve position without first obtaining an operating supervisor's permission.

WEAR THE PROPER GOGGLES WHEN WORKING NEAR THE FLASH OR ARC OF AN ELECTRIC WELDER. WEAR THE PROPER GOGGLES, HELMET OR SHIELD WHEN BURNING OR WELDING.

Pipe rigging and equipment should be thoroughly checked before use.

Electric welding

Eye injuries in welding are very common: Protect your eyes with safety glasses with side shields worn under welders' hoods.

Helpers should also use safety glasses with side shields.

When several lengths of cable are used, use insulated connectors to join them together.

Dispose of weld rod stubs safely. Do not leave them on the floor; they have been the cause of many bad falls.

Allow welding only in designated areas. Control access to areas where welding is being performed.

Underground work is the most hazardous of all. All personnel working underground should be constantly aware of the hazards and take careful precautions.

Almost every activity on the job site requires several basic prerequisites: good lighting, good noise dissipation, good ventilation, and safe working foothold surfaces which are clear of debris. If these criteria are met, in general, there won't be much drastically wrong with your project, because this practice shows the fundamental care which will naturally extend into the rest of the safety program. Job-site safety has become an exact science rather than a guess and a bit of luck. If it is treated with the same respect the other disciplines of work are, then progress will be swift, and some of these horrendous accidents will be avoided.

Carelessness

One major undiscussed subject in safety is carelessness, which is a major cause of lost-time accidents in every type of environment. Let's elaborate on a few precautions which may, by being highlighted here, cause somebody to think twice and thereby avoid a silly, unnecessary situation which could cause an accident.

1. Do not leave tools on ledges to be knocked off onto a person walking underneath.
2. Do not leave welding rod ends all over the floor for people to slip on.
3. Do not leave discarded lumber with nails sticking out, to go through somebody's shoe and foot.
4. Do not leave piles of material stored in a precariously balanced position. You know it could easily fall, but you can't be bothered to fix it properly.
5. Do not leave hand tools on the ground so that, as in the old-time comedy pictures, when the man steps on the blade, the handle comes up and hits him in the face. Of course, if it happened to you, it wouldn't be so funny.
6. If oils and other abrasives or noxious or greasy liquids spill, clean them up. Like a banana skin, a slip on oil can be very painful, and there is no humor in it whatsoever, especially if somebody happens to damage his or her spine. It has happened.

7. Unplug all power tools after use.

8. Isolate all machinery away from the control panel so that somebody cannot inadvertently start the machine.

9. Do not leave electrical leads all wrapped up like spaghetti without proper extension connectors. When somebody goes to straighten out all the knotted cables, at the first touch, he or she could get an electric shock. Keep power cables (and ropes, for that matter) off the ground where possible or sensibly routed so that they are out of people's way.

10. Erect ladders in positions where people don't have to walk underneath them for access or thoroughfare.

11. Broken or faulty equipment should be removed promptly from the field and labeled, avoiding its use by the unsuspecting worker who sees just the tool he or she is looking for, quite unaware that the tool is defective.

Finally, we must address the age-old problem of individual responsibility and the "it's not my job" syndrome which is a major concern and, unfortunately, is prevalent in any walk of life. To illustrate what I mean, here is an anecdote that I came across a while ago which I believe amply explains the problem.

The story of nobody

This is a story of four people named Everybody, Somebody, Anybody, and Nobody. There was an important job to be done, and Everybody was sure that Somebody would do it. Anybody could have done it, but Nobody did it. Somebody got angry about that because it was Everybody's job. Everybody thought Anybody could do it, but Nobody realized that Everybody wouldn't do it. It ended up that Everybody blamed Somebody when Nobody did what Anybody could have done.

The parable of this story is very evident and one that must be addressed by Everybody. When Everybody realizes that safety is Everybody's responsibility, Everybody will be much safer in the workplace.

Drugs and Alcohol Abuse

I have left two topics till the end of this chapter because both are subject to change, and both are politically current enough that many people are not fully aware of what is law and what is not law and what are the minimum requirements which must be followed, as related to construction and industry. The two topics are drug and alcohol abuse and the "right-to-know" laws.

First, let us attempt to deal with drug and alcohol abuse. The government is at present leading the way in trying to design procedures to eradicate drug and alcohol abuse from the workplace, which is a very noble goal.

In the construction industry we have been aware of this problem in the workplace for some time, but we have addressed it in an uncontroversial way. We have maintained the view that while we acknowledge that there is a substance-abuse problem in the construction industry, we do not see it as a very large problem, and mass work force drug and alcohol testing is not going to significantly reduce accidents in all areas of the job site. We believe that the drug and alcohol problem is small because the high degree of coordination, skill, and balance required to perform most jobs in the industrial and construction sectors would result in a much higher accident rate if large quantities of drugs or alcohol were a common influence in this workplace. Further, alcohol and drug traces would be present on accident reports if indeed drug and alcohol abuse was a major contributor to accidents in the construction and industrial workplaces.

However, our contention does not mean we have ignored the problem. In fact, we have recently developed a policy which should be followed and is acceptable to both unions and management. The policy is defined here, and it is being and has been used for years in some form or other by many companies.

Most important, a general substance-abuse statement by management is necessary. This statement could be included in the general company policy statement (see Figure 1.1) and could be worded as follows:

> This company is totally committed to protecting its employees, clients, customers, and the general public while they are in contact with our company job sites or workplaces, from the effects caused by drug and alcohol abuse.
>
> This company has initiated the following procedures and policies to be adhered to on all our job sites and workplaces. It is expected that all employees, contractors, subcontractors, and visitors will follow these policies:
>
> 1. Possessing, furnishing, selling, using, or being under the influence of any illegal drug or other controlled substance or alcohol as defined by federal or state law is prohibited.
> 2. If it is shown that there is reasonable cause to suspect that the above policy has been violated, the company may request that the individual concerned will permit:
>   a. A search of his or her person and/or personal effects
>   b. If necessary, a blood or urine sample from the person, for the purpose of analysis
>   c. Further medical attention to determine the person's ability to work within these safety guidelines
> 3. The company will expel from its job sites or workplaces anybody who there is reason to believe is involved in alcohol or drug abuse.

Both unions and managment have, because of the amount of publicity received by drug and alcohol abuse, spent a great deal of time and effort recently trying to evaluate and quantify the problem and to deal with it in an ever-changing environment. The AFL-CIO, to name but one of several groups, has recently completed a very thorough study, the results of which are to be published shortly and should be well worth reading.* However, the bottom line of all the effort by both employees and unions is that while there is a problem of drug and alcohol abuse (and most importantly, management and unions should work together to address the problem), the problem in the construction and industrial sector seems to have been exaggerated.

---

*The AFL-CIO Washington has published a resolution adopted in January 1987: *Safety, Health and Substance Impairment.*

The "Right-to-Know" Laws

The second topic under public scrutiny these days concerns the "right-to-know" state laws and the "hazard communication" federal law. Many states are adopting right-to-know laws which are designed to make the employees in any situation more aware of the dangers of hazardous chemical substances and to inform them of how to react if contact is made with any of these hazardous chemical substances.

The right-to-know laws may be different from state to state, but they all establish certain minimum requirements for protecting the individual and for reporting hazardous chemical usage and disposal.

It is very important to realize that distinctly apart from these right-to-know state laws is the *federal* hazard communication law. This law includes such requirements as:

> All chemicals defined by the law as hazardous must be accompanied by a MSDS (Material Safety Data Sheet), which must be kept on file as long as the chemical is in the place of work. The Safety Information on the hazardous chemicals used in the place of work should be broadcast to all employees, and these same employees should be trained in "how to react should contact be made with a hazardous substance."
>
> All chemical containers should be clearly, legibly, and indelibly marked with trade name and hazardous contents or category.
>
> All contractors and subcontractors working in the same workplace must exchange MSDS information with the owner when bringing chemicals (for whatever purpose) into the workplace.

Many substances in daily use, not normally suspect, are classified as being of a hazardous nature by federal and state authorities, and the list is not necessarily the same for the federal and state requirements.

These laws and ensuing requirements should be welcomed; they are addressing this subject more effectively than it has ever been addressed before, as this is an area where many, many serious accidents and injuries occur. As more and more research is done, more chemical substances are being found to be hazardous in some form or other. These laws are attempting to bring these facts to the attention of everybody who is exposed to them and to protect employees of companies who deal with any such substances.

Everybody learns of the dangers of acids, fire, and some poisons at a very early age. Many, many more substances are of a hazardous nature than these, not just to the individual but also to the environment. So these laws have been instituted to make the whole scenario—humans, wildlife, and the environment—safer and to improve the general quality of life on this planet.

As a final word on the subject, it is worth pointing out that even if in some cases the state and federal laws seem to be at odds with each other, the most stringent case applies.

Dear Employee,

On November 25, 1983, a federal law was published requiring employers to provide information on any hazardous substance to which you could be exposed in your workplace. This law took effect May 25, 1986.

A Worker and Community Right to Know Law was also published recently. The regulations contained therein are effective already.

Both these laws are designed to make us all more aware of potential hazards within our workplace and community and to establish procedures for treating any conditions arising from contact with these hazards in a safe, speedy, efficient manner to ensure the best possible protection for all personnel in all eventualities.

It is most important that you be familiar with all the hazardous substances in the workplace, their location, and what treatment should be initiated should you come into contact with any of these substances.

Lists of these hazardous substances are posted in each work area. Please become familiar with these lists and the location of these hazardous substances.

Should you come into contact with any of these substances, the procedures listed below should be followed immediately:

1. Determine the substance you have come into contact with by its correct name. This is cearly marked on the container.

2. *Immediately contact your supervisor.*

3. Ask for the MSDS (material safety data sheet) which will define the required treatment. These are available at the manager's office and other supervisor stations including the first aid room.

4. Go immediately to the first aid room. Interim treatment will be available there. Further treatment will be given as required.

We would caution you that many substances in daily use, not normally suspect, are classified as being of a hazardous nature by the federal and state authorities.

You will receive further information and training in connection with this law as required. Should you have any questions, additional information on any part of this law is available at the main office.

We have a very good safety record at this establishment because everybody is continuously aware of the hazards in the workplace. Let us continue this policy by making the effort to participate in these regulations.

Figure 5.1. Example of letter from company to employees concerning hazardous substance exposure.

Figure 5.1 is an example of a letter designed to introduce the subject and training to each employee. Such letters are a basic requirement of the laws. Here are some of the other requirements:

Develop a list of hazardous chemicals in your workplace and their location.

Make sure all chemical containers are clearly, indelibly marked.

Make sure all chemicals on site are accompanied by an MSDS.

Make sure all records are maintained for 30 years.

The federal law took effect May 25, 1986, and it should be regarded as the first giant step to protect the environment from human intrusion, and it is hoped, to protect it for many generations to come.

6

Self-Evaluation of the Safety Program

It is probably now very clear how the safety program works, and it should also be clear that any company can use the program for any type and size of project. It doesn't matter whether the project is fast track, fixed price, cost plus, direct hire, construction management, $0.5 million or $100 million—the system is a sound basis for a construction or industrial site safety program.

However, during the course of the project, the system has to be monitored with checks and balances by the project team to ensure that:

1. The safety program is working properly.
2. Everybody within the program is doing what they are supposed to be doing.
3. The program is continually reviewed, modified, and improved to ensure that it addresses the safety concepts required as applied to each individual contract and project.

There are several simple routines suitable for checking the performance of such a program and the people running it.

Review of Site Inspections

The first group of these checks is identified with the monthly inspection report. This report should give a very detailed, regular indication of faults and failures occurring and the length of time taken to correct any such faults or failures. These checklists should be reviewed. This is a good guideline by which to establish satisfactory or unsatisfactory progress.

But look a little deeper into these reports—they can be very revealing. If the monthly reports consistently show no comments, something may well be wrong! Maybe nobody on the inspection panel is really looking for problems! Do they need further training? If they should show minute or overzealous, unnecessary detail, something also could be wrong! Maybe the committee members are looking at the wrong things or are just being far too critical, which is just as unsatisfactory as not being critical enough! On the other hand, if the monthly report shows faults and the faults are corrected in a timely manner, it can usually be assumed that that part of the system is working just as it should be.

Review of Committee Meeting Minutes

Next look at the safety committee meeting minutes. It is very simple to review them in order to see if the right types of topics are being discussed. These minutes will also elaborate on and indicate the trivialities or justifiable concerns and levels of input from all sides, which is an absolute necessity to ensure a smoothly operating program.

Review of Orientation Sheets

The next group of these internal checks and balances is to assure clear, concise paperwork. This should be reviewed by a responsible person, unconnected with the safety committee.

In most cases construction companies have a corporate safety officer, a contract supervisor, or site administrator who comes to the site at least quarterly and, depending on the type of contract, normally monthly. One of his or her jobs should be to check the safety paperwork to ensure traceability, accuracy, and correct implementation and to ensure that the groups of records are compatible with each other. It is not a big job. If, for example, on a certain date there are two laborers, three millwrights, and one ironworker hired, it is worthwhile checking the payroll register to see if in fact those six persons from three different crafts did arrive on that date. These spot checks are designed to promote confidence in the system. If the check shows everything is correct and everything balances, so much the better. On the other hand, if it proves that only three of the six persons attended an orientation class because only three orientation check sheets can be found, there may be several problems. One obvious reason for the discrepancy could be that communication among site management groups may need to be improved. These quirks always show up somewhere. Find them, recognize the problem and eradicate it. Self-improvement of the program is what you are now achieving. This, of course, is infintely better than having an outside inspection organization like OSHA find these same mistakes, and which often causes a great deal of embarrassment.

This is how they work. They check one set of records against the other to find all the discrepancies and to see how serious the discrepancies are. The more discrepancies, the harder they look; so obviously it is in the site's best interests to do their own reviewing regularly to ensure that this type of embarrassment does not occur.

Another method of arranging internal management checks is to encourage six-month construction reviews performed by the same corporate construction officer which would incorporate an audit of safety records as well as all the normal audit checks. All paper and system deficiencies will become easily recognizable during this series of reviews.

The most important internal reviews, however, are done by the safety committee itself. They should perform many such checks and balances themselves. Some of the important checks and balances for the safety committee to review systematically are described in the following section.

Review of All Accidents

Review all accidents and near-misses and categorize them; review site records against company and national records. Make sure the paperwork is accurate and correctly recorded.

Determine the causes, evaluate particular trends, and attempt to change policy, if necessary, to incorporate the remedies to any negative trends that are identified in the evaluation. Call if assistance from your supervising safety body is required! If they cannot help, your insurance company, OSHA, or the AFL-CIO experts probably can.

Miscellaneous Reviews

1. Review training in all categories, and determine if sufficient training has been achieved or not. Supplement training in weak areas, where applicable.

2. Review monthly site inspections, review each supervisor's daily inspections, look for negative trends, and resolve problems.

3. Occasionally assign one member of the safety committee to check out one section of the paperwork to see if, in fact, it is in good order. Review it against other supposedly compatible sections of paperwork.

4. Constantly review the comments of the suggestion box and others outside the safety committee to see if the general work force has spotted something that has been ignored or overlooked.

5. It is very easy when everything is going smoothly for complacency to set in. This mood must be recognized and reversed. Look for telltale signs. Constant reviews and constant attention to detail and continued attempts by all persons on the safety committee to introduce interesting topics and discussions at all meetings will tend to stop this from happening.

6. Review signs and instructions placed in and around the site. Are they sufficient? If not, add more signs.

7. Review any outstanding enforcement problems and resolve them; suggest improvements to program.

8. All site regulations and rules should also be continually reviewed. On a changing site, it is going to be necessary to rescind some rules, modify others, and institute new ones as the site changes in its phases of construction.

9. Returning to the example given in Chapter 5 where a specific blade failure occurred on a grinding machine: Many such failures occur, but, it is hoped, without personal injury. It is very useful and creates a great deal of confidence with the work force for the company purchasing department, and other related departments, to become involved in evaluating equipment for performance. Maybe the safety committee and engineering and purchasing should all review the bid specifications before going out for bid.

The amount of accidents caused by small tool failure really is alarming. Many companies unwittingly buy cheap imitations of quality tools and do not get either industrial-quality or heavy-duty grade, which leads to early failure. Such false economics are grave mistakes. Heavy-industrial grade tools are imperative on a safe construction or industrial job site.

In conjunction with evaluating hand and small tools, one of the safety committee's regular duties should be to verify when any orders are placed that the right equipment, tool, or machine and the right accessories are provided for the job, thereby ensuring that at least the tools used are the safest available for execution of the task to be performed.

Finally, it is most important to realize that not all the evaluations can be done by the responsible groups on site and within the company. It may be necessary on occasion to enlist the help of outside experts. Industrial hygiene is an important topic and one where additional support may be required. For example, a check on equipment noise levels on the job site is very necessary; are ear defenders required for general use? Besides being dangerous to the ears, high excess levels of noise seriously affect productivity. Another example is to check for noxious fumes and correct storage of chemicals or gases. Such precautions could help avoid serious accidents to the work force. It is very unlikely that there is a person on site able to develop and record these specific requirements. However, there are many experts in these fields—OSHA, the Department of Energy and Resources, and most industrial-sector insurance companies—who will probably not charge a fee at all for these services. If there is a charge, it will assuredly be minimal in normal cases.

All these checks and balances discussed are designed to enable site staff to control their own safety program, their own destiny. After a couple of years' perseverance with the program and its regular reviews, many of these checks and reviews become second nature. It is the same as any other specialized field—it is knowing what to look for and where to look for it. Most importantly, there has to be a good balance. A good safety program has good paperwork and a good clean, safe site. One or the other being good and the other being poor will not do the job.

Consistent reviews will keep everybody on their toes. It will avoid complacency ever getting a grip on the work force, and I am sure you would all agree that this evaluation is better done by you and your colleagues than by your peers. The solution is in your own hands.

Conclusions

It is important at this point in the discussion to delve a little more deeply into a subject touched upon earlier in this chapter, that of using OSHA, your insurance carriers, the AFL-CIO, and other advisory bodies to help you in your quest for the good, all-around safe site and low incidence rate. Oh dear, you may say, that is not a good idea. Well, before you adopt that philosophy, please read on.

Remember, all these organizations are not on this earth solely to enforce rules and regulations. As groups, they have spent a great deal of money and time researching industrial and construction safety. Their advice is good and it is free. They have spent, in some cases, the general public's money achieving this high degree of expertise, and it is there for all of us to use. It seems senseless to me not to take the fullest advantage of it. OSHA and the other organizations have also spent a great deal of time and money researching construction safety programs. All these organizations are sitting on the edges of their seats wanting and waiting to help those who ask.

Bear in mind that there can be no legal action taken against the site concerned if a visitor from one of these groups happens to find a particularly glaring problem or mistake. They will, however, ask you to correct it. They are not going to condemn the site, but they will advise and assist you in improving your program.

A change in attitudes with regard to these organizations will serve to accomplish so much within the industry if it can be achieved. It will encourage mutual respect and a mutual effort to reduce the accident rate in the industrial and construction industries, which is our moral obligation. Let these organizations help you. They can do it very well if you will let them. Let's face it, sometimes we all need constructive advice. Let's start using the people and organizations best equipped to give it.

The Safety Program Audit

All safety control and hazard analysis programs are designed to achieve one end result: to reduce and avoid accidents on the job site and give the whole work force a safe, healthy environment in which to work.

All programs, no matter how basic or how elaborate, have been designed so that there is some record and traceability of the program achieved. Like all records and information generated for any program, there must be method and organization; otherwise, improvements cannot be made or quantified, and the objectives are never reached.

This chapter and the appendix that follows it attempt to provide a universal standard management tool to evaluate and grade any safety control and hazard analysis program to ensure that the program is achieving its goal.

It is important to remember when considering procedures to evaluate and grade a safety program that there are many ways of doing and recording any task or event. No one way is correct. It is important to be tolerant and avoid being critical for criticality's sake. Be positively critical, but look to see if the information required is available, regardless of the format in which it is presented.

The purpose of these procedures is to:

1. Assure the evaluator, and therefore management, that the program is in full compliance with the requirements

2. Assure the evaluator, and therefore management, that the records and information generated serve the purposes for which they are required

3. Ascertain the areas of the program that should be improved or that could be made more efficient

These procedures are set out in a manner that is conducive to one person's doing the complete inspection and evaluation. However, on large projects, for the sake of quantification, over $50 million projects, it may be necessary for the sake of expediency to use two or more persons to perform those same inspections and evaluations. The person or persons selected should be totally divorced from the day-to-day operations of the site, so that there is total impartiality.

Procedure Philosophy

The procedures used here have two major sections: on-site, in-the-field inspections and evaluations and on-site in-the-office inspections and evaluations.

Each section will be broken down into individual subsections or subjects. The idea behind this procedure is to:

1. Allow as much flexibility as possible in the type of persons chosen to perform these duties. They do not have to be very experienced in any or all facets of construction or construction safety policies, as long as they can follow procedures and are able to use good judgment and they have some administrational experience.
2. Allow as much flexibility as possible so that this procedure can be used on any size and type of construction project, regardless of the specialties involved.
3. To develop some kind of checklist to be able to evaluate all aspects of the site safety activities and grade the end results respectively with regard to company history, site history, and the national incidence and severity rates.

Inspection and Audit Checklist Forms

All the checklists to be used during the whole assessment program are contained in the appendix to this book. While the checklists look complicated, they are designed to encompass all types of industrial construction project activities. Obviously, if sections are not applicable, then it should be so stated. Then there will not be a negative impact on the assessment results.

Arrival-on-Site and Entrance Interviews

The first important step upon your arrival is to seek out the site project manager (who knows or should know of your visit at least 48 hours in advance) and to offer an introductory briefing to all concerned parties, i.e., the site manager, labor representatives, safety coordinator, and any other responsible persons involved in the safety program. The introductory briefing should consist of the following:

1. A statement should be made as to the purpose for the visit.

2. The type of inspection and evaluation program should be outlined.

3. A synopsis of the areas to be addressed should be presented.

4. An approximate timetable for the program should be set.

5. It should be decided which of the parties will be involved and with whom you will be directly communicating.

6. It should be explained that there will be a final briefing at the completion of the inspections to which all the same parties will be invited to attend. At this concluding meeting, the scope of the results will be discussed.

7. At the close of the introductory briefing, you and your chosen escort should start the audit proceedings.

On-Site, "In-the-Field" Procedures

These activities will be subdivided into the following topics:

1. Site inspections

2. Employee participation interviews

3. Safety committee member interviews

Site inspections

The site inspection should include all areas of the project and can be performed either with one very detailed tour at the beginning of the program or several tours throughout the program, depending on the size and complicated nature of the project.

It is suggested, however, that the most efficient method is to make several short specific tours into the field, each time concentrating on one aspect, with one introductory orientation trip around the site. With only one long all-encompassing trip, it is very difficult to assimilate all the information required at one time.

In general, the following items should be addressed during the primary tour of the job site. Remember that all evaluations will be marked on the detailed checklist (see appendix) which should accompany each subsequent inspection.

1. Site cleanliness

2. All information and emergency signs properly exhibited with the correct terminology displayed

3. Safety apparatus accessibility

4. Site hygiene—prevailing site conditions, e.g., stagnant water, loose

or waste chemicals, continuous high noise levels, constant dust, and smells.

5. Safety precautions—correct shoring, floor penetrations, barricades, etc.

6. Workers' general attire, including dress, safety accessories, and general appearance.

Taking note of these general attributes will give you a first overall impression of the site. If the site is sloppy in all of the above, then obviously the first impression will not be good. If, however, the site is tidy and complies with all the general requirements, the first impression will be good.

It is very seldom that a clean, tidy site does not have all the attributes of a good program, good record keeping, etc. The reverse is also true, as the site conditions generally reflect the type and quality of contributions of the management and labor forces on the project. Specific inspection details should be addressed with the checklist questions.

Employee interviews

At some point during the subsequent site inspections, it will be necessary to interview several of the craftspersons on site. All the major trades and (if there is enough time) some of the minor trades should be interviewed. This can be done either in the field or in the office. It has been found to be beneficial (for frank responsive answers) to ask the craftspersons planned but spontaneous (effectively spontaneous to the craftspersons) questions while in the field on an inspection, rather than interview them in office surroundings which psychologically are unfamiliar to them.

General questions to the craftspersons are posed to ascertain:

1. Does the individual feel that the job site is safe?

2. Does he or she participate in the safety program in any way, even though he or she is not on the safety committee?

3. Does he or she know the safety committee's function?

4. Does he or she feel that the safety committee is important and effective?

5. Does he or she feel that any more can be done to achieve a safer job site?

6. Is there any on-the-job training done on this site?

The answers to all the above questions concisely evaluate how visible the safety committee is to the work force and how it is doing its job.

The questions can be detailed, but a short, concise interview is probably the most effective and does not promote any suspicions or delay production. It is sensible also to have the site escort present during the interviews so that all the replies and nuances can be evaluated jointly later.

Safety committee interviews

These interviews can be conducted on site, but some time has to be spent in the offices. The questions should also be a little more detailed. The purpose of these interviews is to judge, with the help of the safety committee meeting minutes, if the committee members are being effective in organization and administration and how to improve the situation if the system is not effective.

So a great deal more time is needed to review the meeting minutes and address a series of questions to the committee members individually, both in the field and in the offices.

The following is a selection of the types of questions which may lead to illuminating discussions with the individual members:

1. "Briefly, run me through, in your own words, a typical safety committee meeting. What happens?"
2. "Do the same people speak all the time, or are all the committee members putting in their suggestions?" (If the former, it may be that only a few people are dominating the committee. This is not the ideal situation.)
3. "Does the safety committee always use the same people on the monthly site inspections?" (Inspectors should be rotated.)
4. "Do you enjoy being on the safety committee?"
5. "What was the last accident you all discussed at a safety meeting?"
6. "Do you participate in daily site inspections? Do you have a notebook in which to record problems? Can I see your inspection notebook?"

Answers to all these questions will very easily determine the efficient safety committee from the committee that is not planning and controlling safety on the job site.

The following steps in the audit take place over several hours, and they are very important. It must be noted that doing these interviews intermittently over several hours, instead of one after the other, is refreshing and tends to keep the mind focused on the overall picture rather than on the constant involvement in minute detail which may tend to throw the evaluator off course.

When you are satisfied with all the information you have accumulated, the checklist questions can now be completely answered.

On Site, "In the Office"

The second area of inspection and evaluation is the paperwork and administration side of the safety program. No matter which safety program is functioning on site, there is a certain amount of "reporting" paperwork which is mandatory:

1. OSHA 200 log (filled out monthly and displayed annually)
2. Accident register
3. Indoctrination forms for employees
4. Accident reports, compensation reports, return-to-work slips
5. Accident investigation reports
6. Crane inspections
7. Monthly inspection reports
8. Hazardous chemical log and records

These are the basic forms necessary to run a job site at any level because the state and/or the insurance company that insures your particular site demands that this paperwork be available; without this paperwork insurance policies will not be approved or upheld.

In addition to the eight reports above, there are others which make administration of a safety program that much easier. These are listed below:

1. Misconduct reports and actions
2. Near-miss evaluation reports
3. First aid and/or CPR personnel log
4. Toolbox meeting reports
5. Fire extinguisher log
6. Safety committee meeting minutes
7. Inspection reports
8. Details of any emergency and evacuation plan

Any half-efficient construction site needs records on all sorts of controls and activities, not just safety. Otherwise nobody is able to assess productivity, percent completion, and expenditure, all of which have a major impact on the cost of the project and, therefore, profit and loss.

The effective records of the safety department enable very quick evaluation of costs of insurance and claims to date and they also supplement other valuable information in giving management a complete cost picture. A lack of records indicates that management is not getting the complete picture and therefore has no confidence in the job site or its progress.

Each one of the sixteen groups of records should be inspected if they are relevant. If not, inspect what is available! Obviously, the primary eight records should exist in some form or other.

The inspections should show the following in each area of the records:

1. Current entries:
    a. One week's lapse is acceptable.
    b. Two weeks shows cause for concern.
    c. Three weeks is unacceptable.
2. Compatibility:
    a. The number of indoctrination forms should match the current work force.
    b. Dates should be spot-checked for accuracy.
    c. The correct literature, rules, and regulations, etc. should have been distributed and recorded in the proper place.
    d. The accident log should be compatible with the OSHA 200 form.
3. Thoroughness—recording and filing information:
    a. Correct information is in the correct file.
    b. Chronological reports are in order, and all present.
    c. Meeting minutes are correctly distributed.
4. Correct review of any accident and correct documentation and possible changes as a result of the accident

All these results should now be documented on the checklist forms as shown in the appendix.

Remember, nobody is perfect; therefore, odd mistakes are going to be uncovered. This is not a major problem, *nor* should it be treated as such. Point out the mistakes, and have them corrected. Should, however, the mistakes be prolific, then there is cause for concern, which should be voiced. (See the discussion on exit interviews later in this chapter.)

Finally, before the paperwork and administration inspection is complete, an evaluation should be made of the corporate management commitment. That is relatively easy to assess, with the following suggestions:

1. Is there a company statement on safety on the job site?
2. Have there been visits by corporate personnel with any orientation toward safety?
3. Are there good labor-management relations on site?

If the answer to all these is yes, then there is good corporate management commitment!

Chances are if there is this type of evaluation being done in any way, shape, or form, then there is good corporate management commitment, but it still has to be reviewed.

However, if the answer to all the above is no, then make doubly sure the inspections and evaluations are handled in a positive, professional manner, sticking only to the facts with no conjecture. Then maybe corporate management may well be convinced that the need for commitment exists and the whole exercise will provide a positive attitude for the future.

Remember, being fresh to the site, you will be able to evaluate much more quickly and effectively than the incumbent site staff. Be sure to use the site staff as much as possible in your inspections so that they see the whole spectrum with different eyes; at the same time this practice will go a long way to ensuring that improvements occur prior to the next visit because the site staff will see the project through your eyes rather than their own.

Site Inspection Evaluation and Exit Interviews

The whole site inspection checklist should now have been completed. The whole site safety program now needs to be evaluated.

The first 21 topics and questions in the appendix should be generally the same as those used for each monthly site inspection by the safety committee. The reason for this is that the same items are covered consistently, thereby eliminating the arguments such as "you have changed the rules for this inspection" or "the inspection is being done in a different way." Everthing in the inspections is constant, and therefore the checklist is a very good tool to use for evaluations and in assessing change in the program's effectiveness.

The last four topics and questions cover general administration, without which there would be no documentation and, therefore, no meaningful safety program.

Evaluation

Using a points system is a matter of personal company preference. However, it has been the experience of this writer that points systems do not really convey accurately the end result for the following reason: If you give each question a point value, the overall program loses its importance in some very visible areas. For example, there are some items being audited that cannot be inferior without the program's being completely in jeopardy, but a points system would not reflect this.

Therefore, the following ideas on program evaluation do not include points; instead, the evaluation depends purely on an analysis of each negative answer.

Evaluation of each negative answer should take place after the whole checklist is completed. The evaluation should ask one question, and one question only.

Can the safety program function with this item missing? If the answer is yes, then the program is working and acceptable. If the answer is no, then the program is not totally acceptable.

Now, there are degrees of acceptability insofar as the program may not be as "all encompassing" as it should be even though it is still working. However, assuming that each individual site has the minimum program introduced to the site, then all no answers are to be reviewed as an important negative trend which needs correction. Correcting negative trends, speedily, is of prime importance.

The accident record of a site, which has been "audit" evaluated as being very poor in its safety program, could still have a very low accident rate. However, surely by ignoring the proven evaluation procedures and recommendations that this inspection critique suggests, the site is leaving itself highly exposed to the hazards and the "opportunity" for accidents to occur.

This is the fundamental reason for this audit: to ensure that as many of these opportunities are nullified as much as possible, to avoid an accident. Of equal importance and maybe as persuasive an argument is the fact that having a bona fide safety program will reduce insurance premiums.

In sum, therefore, the program audit is either acceptable, good, or poor. Acceptable means the no's on the checklist do not jeopardize the program but need to be addressed quickly. "Good" means minor changes may need to be made but the whole program is being followed and documented properly. "Poor" is anything else, which is unacceptable!

Site visit summary and exit interview

Now comes the tough task of reviewing this inspection and evaluation with the same people who participated in the entrance interview. Whatever the results, this should follow a pattern as detailed below or similar, in order to encourage the positive responses of the site as opposed to negative responses:

1. Give the safety coordinator and site manager as much notice as possible to reconvene all parties for the exit interview.
2. *a.* Get the site manager and the safety coordinator to one side and give them a brief synopsis of the audit findings.

    b. Arrange for all parties at the exit interview meeting to have a copy of the inspection checklist.

    c. Make sure that the site manager and safety coordinator sign the original as having read and received one copy.

3. Even if the program and inspection and evaluation are poor or worse, start the proceedings of the exit interview with:

    a. Thanks for the assistance and cooperation by all parties involved

    b. A list of the good points of the visit and the audit

    c. A brief list of the items which did not conform to the program

    d. A brief list of simple suggestions as to how these negative items can be improved immediately

4. *a.* Explain the rough contents of the written report which will be presented to the auditor's supervisors.

    b. Explain that a copy will be sent to the site manager and safety coordinator.

    c. Open the exit interview proceedings to comments from the site manager and then to a brief question-and-answer session.

    d. Then excuse yourself.

Remember manners, etiquette, civility, and sympathy are most important in the whole procedure, from arriving on the site to leaving the site. It is absolutely necessary, and cannot be stressed too highly, to encourage improvement in site affairs by a positive attitude rather than an abrasive attitude.

Site Evaluation Report

The site inspection and evaluation program is complete, and all that remains is to report to the corporate office with the evaluation results and the site visit report.

There is no merit in just sending in the checklist and leaving the readers to make up their own minds. This is dangerous. Everything can be blown out of proportion or alternatively can create negative response by management. It is important to qualify the report with positive thoughts and to highlight the positive actions. In all cases this makes the "bottom line," even if it is miserably poor, that much easier to absorb. Most important, if the end result of the audit is accepted, then more positive actions, and therefore positive site safety attributes, will be established.

Efforts have to be made. Goals have to be accomplished. Progress has to be made. This is all totally positive. Everybody must contribute. Then the absolute objectives will be achieved: a safer project resulting from a fair, honest assessment of the safety program.

A Typical
Audit Inspection Checklist

Date: _____ Location: _____

Subject	Yes	No	Remarks
1. GENERAL (Conditions of Job Site)			
a. Posting OSHA and other job-site warning posters			
b. Records of safety meetings up to date			
c. Toolbox meeting minutes up to date			
d. Availability of first aid equipment and supplies			
e. Job-site OSHA documentation up to date			
f. Emergency telephone numbers such as police department, fire department, doctor, hospital, and ambulance posted in the proper places			
g. General cleanliness of working areas			
h. Regular disposal of waste and trash in the proper manner			
j. Passageways and walkways free from obstruction			
k. Adequate lighting of all work areas			
l. Nails removed or bent over on all lumber and fixtures			
m. Proper chemical waste containers used for disposing of segregated waste in the appropriate manner			
n. Sanitary facilities adequate and clean; more than one unit for 20 people			

(Continued)

Subject	Yes	No	Remarks
o. Good supply of clean drinking water			
p. Sanitary drinking cups available			
q. Correct head protection used by all employees outside offices			
r. Eye protection readily available			
s. Face shields readily available			
t. Respirators and masks readily available			
2. ELECTRICAL INSTALLATION a. Adequate wiring, well insulated			
b. Fuses provided			
c. Ground fault circuit interrupters in place			
d. Electrical dangers posted			
e. Proper fire extinguishers provided			
f. Terminal boxes equipped with required covers			
3. HAZARDOUS CHEMICALS a. Fire hazards identified			
b. Proper types and number of extinguishers easily accessible			
c. All containers clearly, indelibly identified			
d. Storage practices compatible with safety regulations			
4. WELDING AND CUTTING a. Screens and shields used for personal protection			
b. Protective clothing available and used			
c. All equipment in safe operating condition			
d. Power cables undamaged and protected			
e. Fire extinguishers of proper type readily available			
f. Regular inspection for fire hazards			
g. Flammable materials well ventilated and well away from hazards			
h. Gas cylinders secured upright			
i. Gas lines in good condition and protected from hazards			
j. Cylinder caps used in the correct manner			
5. TOOLS (Hand Tools) a. Employee-owned tools checked on and off site and inspected			

(Continued)

Subject	Yes	No	Remarks
b. Damaged tools repaired or replaced promptly			
c. Proper tool being used for the job			
d. Safe storage, adequate facility for carrying safely			
TOOLS (Power Tools) a. Proper grounding applicable for each set of tools			
b. Proper training carried out			
c. All mechanical safeguards in working order			
d. Right tool being used for the job			
e. Tools neatly stored when not in use			
f. Tools and cords in good condition			
g. Good housekeeping for tools and auxiliaries			
6. MACHINES AND EQUIPMENT a. Safety goggles or face shields used where required			
b. Tools used where recommended			
c. Proper training for machines and equipment is carried out			
d. Compliance with all local laws and ordinances			
e. All operators qualified to operate respective machines			
f. Machines in good working order			
g. Machines and equipment protected from unauthorized use			
7. GARAGES AND REPAIR SHOPS a. No fire hazards			
b. Fuels and lubricants dispensed cleanly and safely			
c. Good housekeeping throughout shop			
d. Lighting adequate for safe working			
e. Carbon monoxide dangers identified			
f. All fuels and lubricants stored in proper containers in properly ventilated area			
8. FIRE PREVENTION a. Phone numbers of fire department adequately displayed			
b. Records exist of fire instructions, planning, and training given to employees			

(Continued)

Subject	Yes	No	Remarks
c. Fire extinguishers checked and documented			
d. "No Smoking" displayed where necessary			
e. Fire hydrant access to public thoroughfare open			
9. HEAVY EQUIPMENT a. All lights, brakes, warning signals operative			
b. Regular inspection and maintenance shown			
c. Lubrication and repair of moving parts recorded			
d. Roadways well maintained, accessible, and clear of interferences			
e. When equipment not in use, it is unable to be used by unauthorized personnel			
f. Wheels chocked when necessary			
10. MOTOR VEHICLES a. Local and state vehicle laws and regulations observed			
b. Qualified operators			
c. Regular inspection and maintenance shown			
d. All brakes, lights, and warning devices operative			
e. All glass in good condition			
f. Weight limits and load sizes controlled			
g. Personnel carried in a safe manner			
h. Fire extinguishers installed where required			
i. Reversing alarms provided			
11. SCAFFOLDING a. Scaffold tied into the structure correctly and safely			
b. Scaffold walkways free of debris, snow, ice, and grease			
c. All adjacent workers protected from any falling objects			
d. All connections checked for rigidity			
e. Guardrails, intermediate rails, and toeboards erected and sufficient			
f. Erected in accordance with the regulations and checked			

(Continued)

Subject	Yes	No	Remarks
12. LADDERS a. Properly secured to prevent slipping or falling			
b. Ladders inspected regularly and in good condition			
c. Handrails extend 36 inches above top of each landing			
d. Metal ladders not used around electrical hazards			
e. Ladders not painted			
f. Safety feet in use			
g. Stepladders fully open when in use			
h. Rungs or cleats not over 12 inches on center			
i. Job-built ladders constructed of sound materials and meet regulations			
13. BARRICADES a. Floor openings planked over or adequately barricaded			
b. Adequate lighting provided in all work areas			
c. Traffic controlled to all work areas			
14. MATERIAL HANDLING AND STORAGE a. Materials stored or stacked safely			
b. Fire protection available			
c. Stacks on firm footings not stacked too high			
d. Clear passageways around material			
e. Proper number of workers for the working conditions			
15. EXCAVATION AND SHORING a. Excavations barricaded properly			
b. Correct shoring used for soil and depth			
c. Adjacent structures properly shored			
d. Equipment a safe distance from edge of excavation			
e. Proper ladders provided where needed			
16. STEEL ERECTION a. Hard hats, safety shoes, gloves being used			
b. Taglines and safety belts in use			
c. Floor openings covered and barricaded adequately			

(Continued)

Subject	Yes	No	Remarks
d. Ladders, stairs, or other safe access provided			
e. Hoisting apparatus checked regularly			
f. Safety nets or planked floors in use			
17. HOISTS, CRANES, AND DERRICKS a. Outriggers used when required			
b. Cables and sheaves inspected regularly			
c. Loading capacity at lifting radius below the limit for the unit			
d. Signalers used when necessary			
e. Equipment firmly supported when lifting			
f. Signals understood by operator			
g. Power lines inactivated, removed, or at a safe distance from all lifting exercises			
h. All equipment properly lubricated and maintained			
i. Inspections and maintenance logs maintained			
18. CORPORATE ADMINISTRATION a. Is there a general policy statement by senior corporate management with regard to safety and hazard control on site?			
b. Is this policy: (1) Incorporated in employee indoctrination? (2) Posted in a conspicuous place?			
c. Are there conspicuous official notice boards on site?			
d. Are there written safety regulations: (1) On site? (2) Are they incorporated in indoctrinations? (3) Are they explained to all employees?			
19. SITE ADMINISTRATION a. Does the safety committee meet weekly?			
b. Are the committee meetings documented properly?			
c. Is there written documentation of accident investigation and analysis with corrective measures?			
20. SITE RECORDS a. Has all site training been documented? (1) First aid? (2) CPR? (3) Hazard recognition? (4) Craft skill training, e.g., Hilti Gun operation?			

(Continued)

Subject	Yes	No	Remarks
b. Accident records: *Frequency:* On site [ ] Company overall [ ] National average [ ] *Severity:* On site [ ] Company overall [ ] National average [ ] Are the site figures the lowest?			
c. Is the accident register up to date and accurate?			
d. Are the indoctrination records up to date and accurate?			
e. Is there a documented, planned preventive maintenance program?			
f. Are the correct hazard data sheets available for all hazardous chemicals on site?			
g. Are the monthly inspection reports current?			
h. Is there action taken on the monthly report negative comments?			
i. Are safety equipment warranty and specification records current?			
j. Are tool warranty and specification records current?			
21. GENERAL SITE CRITERIA a. Do employee interviews show participation and interest in the safety program?			
b. Do safety committee interviews show interest and competence in the program?			
c. Is safety and emergency apparatus easily accessible, e.g., breathing apparatus, straight jacket, personnel lifting crib, axe, or ear plugs and/or defenders?			
d. Are the general workers on site dressed acceptibly for the job, e.g., hard hats, working shoes, or safety glasses? Are workers dressed neatly?			
e. What medical facilities are available on site? First aid on site? SRN within 5 minutes of project offices? Accident plan?			
f. Is there a site emergency plan?			
g. Is there a site evacuation plan?			

SUMMARY
Last site inspection and evaluation: Good [ ] Fair [ ] Poor [ ]
Was this inspection within one year of this date? Yes [ ] No [ ]
This site inspection and evaluation is: Good [ ] Fair [ ] Poor [ ]

References

AFL-CIO: *Safety and Health Competent/Qualified Person*, pamphlet.

Business Round Table: *Improving Construction Safety Performance*, paper published by Business Round Table, 200 Park Avenue, New York, NY, 10166, January 1982. This paper deals with safety program costs.

Firenze, R. J.: *Guide to Occupational Safety and Health Management*, Kendall/Hunt Publishing Company, Dubuque, Iowa, 1973.

Firenze, R. J.: *The Process of Hazard Control,* Kendall/Hunt Publishing Company, Dubuque, Iowa, 1978.

Hopf, Peter S.: *Designer's Guide to OSHA*, McGraw-Hill, New York, 1982.

National Safety Council: *Accident Prevention Manual for Industrial Operations*, 6th ed., National Safety Council, Chicago, 1969.

Peterson, Dan: *The OSHA Compliance Manual*, rev. ed., McGraw-Hill, New York, 1979.

Index

Accidents:
  common causes of, 37, 73–75
    to boilermakers, 75
    to carpenters, 73
    to electricians, 74
    to ironworkers, 74
    to laborers, 73
    to masons, 75
    to millwrights, 74
    to operating engineers, 74
    to painters, 74
  prevention of, hints on, 78–82
  reviews of, 66
  sources of, 76–77
Activity, definition of, 42
Activity hazard analysis (AHA), 30–31,
  42–44, 64, 67
AFL-CIO, 87, 93, 95
Audit procedures, 67, 97–113
  employee interviews, 100
  office inspections, 102–104
  safety committee members' interviews,
    101
  site inspections, 99

Barricades, 79
Boilermakers, 75
Bureau of Statistics, 36, 59

Carpenters, 73
Communication, 2, 16
  safety committee and, 51
Concrete work, 80
Construction:
  phases of work, 11, 42–55
    commissioning (plant start-up), 51–
      54, 71

Construction, phases of work (*Cont.*):
    construction, 49
    demobilization, 55
    mobilization, 47
    schedules, 49, 50, 58
    start-up, 51–54, 71
    planning a safety program for, 40, 45,
      57
Corporate policy statement, (figure) 5
CPR training, 64, 66, 78
Craft spokespersons, 7–9
Craft training cards, 7–8
Cranes, 80

Daily job-site inspection, 33–35
Definitions, 42
Documentation, reporting, 18–31
  (*See also* Forms)
Documentation review, 91–95
Drug and alcohol abuse, 85–87

Electricians, 8, 74
Employee interviews, 100
Employee participation, 39
Equipment, safety, 64, 65, 68, 69, 77–78
Evaluation of safety program, 91–95,
  104–106
  conclusions, 95
  review, 91–95

Federal hazard communication law, 88–
  89
Filing system, (figure) 18
Floor openings, (figure) 6
Forms, 16–31, 107–113
  Activity Hazard Analysis Form
    (Form 1.8), 30, (figure) 43

Forms (*Cont.*):
Activity Hazard Analysis Worksheet (Form 1.9), 31, (figure) 44
Crane Operator's Daily Inspection Report (Form 1.6), 29
Employee Safety Orientation (Form 1.1), 19
Fire Extinguisher Inspection Report (Form 1.5), 28
Monthly Construction Inspection Checklist (Form 1.4), 22–27
Safety Deficiency Report (Form 1.7), 30
Supervisor's Report of Accidents and Investigations (Form 1.3), 21–22
Weekly Toolbox Meetings (Form 1.2), 20

G's in safety, 39
General contractor, program organization for, (figures) 13–14

Hints on accident prevention, 78–82
Historical data hazard analysis, 36–37

Inspections, 22–27, 33–36, 99, 102–104
Inspectors, site, 67, 91–96
Ironworkers, 65, 68, 74

Job hazard analysis (JHA), 40–41
Job-site inspections, 22–27, 33–36, 99, 102–104
Job-site inspectors, 67, 91–96
Job-site rules, (figure) 6

Keys to safety:
planning, 45–55
tagging system, 51–55

Labor unions, 87, 93, 95
Laborers, 8, 73
Laws:
federal hazard communications, 88–89
OSHA, 1–2, 36, 88
right-to-know, 88–89
Lost-time accidents, 37, 73–75
(*See also* Accidents)

Masons, 75
Material safety data sheet (MSDS), 88–89
Meetings:
safety committee, 64, 68, 92
toolbox, 9–10
Millwrights, 74
Module, definition of, 42
Monthly job-site inspections, 22–27, 35–36

National Safety Council, 36
NIOSH (National Institute of Occupational Safety and Health), 36

Operating engineers, (figure) 8, 74
Operations overlaps, 34
OSHA (Occupational Safety and Health Administration), 36, 59, 88, 93, 95
construction regulations, 1–2, 36, 88
standards, 2

Participation:
on safety committee, 10
by workers, 39
Personnel:
job-site orientation for, 4, 5
orientation checklist for, 5–6
Pipe fitters, 81
Planning:
bid stage of, 45, 46
for cleanliness of site, 40
by professionals, for safety program, 50
of purchasing on site, 39, 65–66
by safety committee, 10–16
of safety strategies, 45
Plant commissioning, 51–54, 71
Prevention of accidents, 78–82
Program evaluation, 91–95, 104–106
conclusions, 95
review, 91–95
Project bidding, 45, 46

Qualifications:
of committee members, 7–9
of safety coordinator, 3

Reporting documentation, 18–31
(*See also* Forms)
Reporting system, 16–18
Responsibilities:
  of committee members, 7 9
  of safety supervisor, 3–4
Right-to-know laws, 88–89

Safety:
  cartoons to promote, 75–86
  and hints on accident prevention, 78–82
  keys to: planning, 45–55
    tagging system, 51–55
  slogans for, 39, 40, 51, 73, 85
  (*See also* Toolbox topics on safety)
Safety committee:
  communication by, 51
  duties of, 12–13
  meetings of, 64, 68, 92
  organization by, 13–16
  participation on, 10
  planning by, 10–16
  reporting system and, 16–31
  responsibilities of, 7–9
  size of, 14–16
Safety control systems, 16–31
Safety equipment, 64, 65, 68, 69, 77–78
Safety hazard analysis, 37–38
Safety hazards, 79–89
Safety program:
  activities to consolidate for, 57
  advantages of, 39
  audit of, 67, 97–113
  costs of, 45, 46, 58
  evaluation of, 91–95, 104–106
  forms for, 16–31, 107–113
  functions to consolidate for, 57
  planning by professionals for, 50
  reporting in, 16–31, 107–113

Safety program (*Cont.*):
  slogans for, 39, 40, 51, 73, 85
  supervisor of, 3–4
  tagging procedure for, 51–55
Sample forms, 17–31, 107–113
Scaffolds, (figures) 6, 63
Schedules, construction, 49, 50, 58
Site demobilization, 55
Site inspectors, 67, 91–96
Site mobilization, 47
Site orientation review, 92–93
Special job-site rules, (figure) 6
Subcontractors, 47
  contract conditions for, 48, 49

Task, definition of, 42
Toolbox meetings, 9–10
Toolbox topics on safety, 60–63
  back injuries, (figure) 62
  electrical wiring, (figure) 62
  eye protection, (figure) 61
  floor openings, (figures) 6, 61
  hard hats, (figure) 60, 65
  ladders, (figure) 60
  makeshift scaffolds, (figure) 63
  reducing injuries, (figure) 63
  safety nets, 65
  scaffolding, 6
Truck drivers, 7

Unions, 87, 93, 95

Wall openings, (figure) 6
Weekly job-site inspections, 35
Welding:
  electric-arc, 82
  gas, 79–81
Worker participation, 39

ABOUT THE AUTHOR

David Goldsmith became a marine engineer in the 1960s and has been involved in industrial safety programs for many years. As president of Goldsmith Engineering Services, he develops safety and inspection programs and provides training in safety to all walks of industry in the eastern United States. He is also a consultant to the federal Voluntary Safety Protection Program instituted by OSHA and is a consultant to the Pennsylvania government on hazard-protection policies.